ARSON RESOURCE DIRECTORY

FEDERAL EMERGENCY MANAGEMENT AGENCY

UNITED STATES FIRE ADMINISTRATION

This publication was supported by the Federal Emergency Management Agency/United States Fire Administration grant EMW-92-C-3954. Its contents do not necessarily represent the policy of the funding agency. Moreover, the opinion and views expressed herein should not be construed to represent the policy of EEI, which produced this publication.

TABLE OF CONTENTS

INTRODUCTION

Background

Public Law 95-422, October 5, 1978, established a Federal focus within the Federal Emergency Management Agency (FEMA) on the serious national problem of arson. Prime coordination responsibility within FEMA was given to the U.S. Fire Administration. The purpose of this responsibility is to provide technical assistance, do research and development, and assist States and localities to develop policies which will help reduce, control, and prevent their incidents of arson.

The second edition, which updated and expanded the first Directory, was published in January 1982, and a third edition was published in 1988.

Purpose

The Arson Resource Directory is intended to provide an explanation and identification of organizations and individuals who are concerned with arson prevention and control. This Directory is designed to help you contact resources that will assist you in coordinating your efforts with others.

Design

The Directory is divided into three parts. **Part I** contains the major programmatic areas, across the Nation, concerned with arson prevention and control. **Part II** identifies miscellaneous groupings of Federal, public, and private resources other than those falling into the programmatic areas. **Part III** is an alphabetical index of organizations, programs, and resources listed in the Directory. The **Table of Contents** identifies major organizational Parts, Sections, and Subsections. The **Index** identifies programs, organizations, and persons alphabetically and indicates page numbers of location.

PART I

A. Management of Arson Prevention and Control Activities and Resources

MANAGEMENT OF ARSON PREVENTION AND CONTROL ACTIVITIES AND RESOURCES

This country is combatting arson using various management approaches, agents, and agencies. The prevention and control of this complex crime require a number of management strategies, including the Arson Task Force concept with its State, county, and metropolitan units.

This section also lists other agencies and entities responsible for management approaches, Data Resources, Arson Information Management Systems, and Public Education Programs.

ARSON TASK FORCES

The arson task force is a management system for the purpose of developing and implementing strategies to control and prevent arson. Its concept is simple: mobilize public and private resources; identify and coordinate responsibilities; set policy or guidelines; and integrate efforts of agencies, groups, and persons who are or should be involved in an organized strategy for implementing arson prevention and control. Arson task forces may fall into two general categories:

- Legal or quasi-legal constituted bodies

- Private, advisory committees, usually initiated or supported by the insurance industry.

Responsibilities of the Arson Task Force

The arson task force has two major areas of responsibility. The first is policy setting, the second is program implementation.

Policy-Setting Responsibilities

1. Defining the community's arson situation by:

   - Identifying the problem
   - Identifying the contributing factors
   - Identifying and evaluating existing programs, if any
   - Identifying available resources.

2. Setting goals and objectives such as:

   - Classifying the magnitude of the arson problem
   - Promoting public awareness
   - Increasing ratio of convictions for total number of arson incidents.

3. Setting policies, establishing priorities, and selecting programs such as:

   - Establishing training for all line firefighters, fire officers, and police officers

   - Encouraging high professional standards for arson investigators and others concerned with arson prevention and control

- Defining responsibilities and jurisdictions for investigation of suspected incendiary fires

- Fostering and improving communication, coordination, and cooperation among fire departments, law enforcement agencies, the insurance industry, and other public and private groups

- Establishing tip&r/hotline program

- Establishing public awareness and media campaigns

- Establishing liaison with the legislators to promote legislation which will provide disincentives to arsonists.

4. Approving and implementing programs such as:

- Identifying sources of funding support
- Obtaining commitments from member agencies
- Identifying responsible agency for specific programs
- Implementing programs.

5. Evaluating and revising, as necessary, mission and goals:

- Assessing program objectives
- Determining program effectiveness and resource utilization
- Examining, as necessary, program alternatives.

Program Implementation Responsibilities

1. Forming an arson control unit

2. Setting up mechanisms for arson data analysis and an arson information management system

3. Increasing public awareness of the arson problem

4. Instituting tipster/hotline program

5. Obtaining community participation

6. Establishing training in detection, investigation, and prosecution

7. Instituting a juvenile firesetters counseling program.

Representative membership in arson task forces usually includes:

Statewide Arson Task Forces	County/Metropolitan Arson Task Forces
The Governor's Office	Office of the Mayor/City Manager
State Legislature	City Council/County Supervisors/
Division of Criminal Justice	Selectmen
State Fire Marshal	Fire Chief/Fire Marshal
Department of Insurance	Police Chief/Sheriff
Division of State Police or Public	Office of the Prosecutor
Safety	Office of Buildings/Codes/Records
State Fire Training Director	State/Federal Forestry Service
Division of Community Affairs	Insurance Industry
Paid Fire Departments	Civic Organizations
Volunteer Fire Departments	Chamber of Commerce
Mayor	Community-based Organizations
Insurance Industry	Media
Police Chiefs/Sheriffs	Public at Large
County Prosecutors	
State/Federal Forestry Service	
Public at Large	

STATEWIDE ARSON ORGANIZATIONS

The Offices of the Governor in several States have been urged to establish arson task forces because of the alarming increase in arson and the accompanying heavy losses of life and property. These quasi-legal task forces examine arson problems and develop statewide strategies for arson prevention and control. Technical assistance by the Federal Emergency Management Agency (FEMA) and initial financial support from the Law Enforcement Assistance Administration (LEAA) in 1979-1980 aided in implementing this concept.

STATEWIDE ARSON ORGANIZATIONS

State	Contact	Address
Alabama	Paul Durham Chairman	Alabama Arson Prevention Task Force c/o State Farm Fire and Casualty Company P.O. Box 94097 Birmingham, AL 35220 (205) 731-1205
Alaska	Jack R. McGary State Fire Marshal	5700 E. Tudor Road Anchorage, AK 99507-1225 (907) 269-549 1
Arizona	Barbara S. Gelband Assistant Attorney General	c/o Arizona Attorney General's Office 402 W. Congress Street Suite 315 Tucson, AZ 85701 (602) 628-6504
Arkansas	Ray E. Careahan Secretary/Treasurer	State Fire Marshal Arkansas Arson Advisory Committee 3 Natural Resource Drive, P.O. Box 5901 Little Rock, AR 72215 (501) 221-8258
Colorado	Shannon Kelly	Colorado Advisory Committee on Arson Prevention c/o Western Insurance Information Service 6565 S. Dayton Street, Suite 2400 Englewood, CO 80111 (303) 790-0216
Connecticut	Howard Nielsen Manager	Anti-Arson Committee of Connecticut Fair Plan 224 Pitkin Street East Hartford, CT 06128-0978 (203) 528-9546

State	Contact	Address

Florida	Guy E. Burnette, Jr.
	President
		Florida Advisory Committee on Arson
		   Prevention
		Butler, Burnette & Pappas
		Bayport Plaza
		Suite 1100,6200 Courtney Campbell
		   Causeway
		Tampa, FL 33607
		(813) 281-1900

Georgia.	Ray Farmer Chairman	Georgia Arson Control Program, Inc. c/o American Insurance Association Atlanta Plaza, 950 E. Paces Ferry Road Suite 2240 Atlanta, GA 30326 (404) 261-8834
Illinois	Stephen P. Eisenberg Chairman	Illinois Advisory Committee on Arson Prevention c/o Leany and Eisenberg Limited 29 S. LaSalle Street Suite 300 Chicago, IL 60603 (3 12) 368-4554
	Carol Jackson Office Manager	Illinois Fire Inspectors Association 200 E. Wood Street, Room 256 Palatine, IL 60067 (3 12) 439-6635
	Frank Rice General Manager	Arson Advisory Board Illinois Fair Plan Association 2 Prudential Plaza, Suite 1400 Chicago, IL 60601-6713 (312) 861-0385
Indiana	Tom McKeon President	Indiana Arson and Crime Associates, Inc. P.O. Box 2282 Indianapolis, IN 46206 (3 17) 842-0096

STATEWIDE ARSON ORGANIZATIONS (Continued)

State	Contact	Address
Iowa	Lloyd Anderson Chairman	Arson Alert Program of Iowa c/o Iowa Fair Plan 6967 University Avenue Des Moines, IA 50311 (515) 255-9531
Kansas	Charles Shrader Chairman	Kansas Arson Advisory Committee, Inc. c/o Upland Mutual Insurance Route 2 Chapman, KS 67431 (9 13) 762-4324
Louisiana	Jack Manifold Manager	Louisiana Property Insurance Association P.O. Box 60730 New Orleans, LA 70160 (504) 831-6930
Maine	Thomas E. Gagnon President	Maine Arson Information Award Program, Inc. Mutual Fire Insurance Company P.O. Box 537 Saco, ME 04072 (800) 635-0915
Maryland	Ed Dillon Chairman	Arson Control Association of Maryland c/o State Farm Insurance Company 2221 N. Broad Birch Drive Silver Spring, MD 20904 (301) 622-7377
Massachusetts	Jennifer Mieth Program Manager	Suffolk County Arson Prevention Program 1010 Commonwealth Avenue Boston, MA 02213 (617) 566-4500, Ext. 316
Michigan	Conard Golemba Chairman	Michigan Arson Committee c/o Auto Club Insurance Group 1501 Washington Boulevard Detroit, MI 48226 (3 13) 237-5557

STATEWIDE ARSON ORGANIZATIONS (Continued)

State	Contact	Address
Mississippi	Millard D. Mackey Chief Deputy Fire Marshal	Mississippi Arson Advisory Committee c/o Office of the Fire Marshal 416 Woolfolk Building P.O. Box 22542 Jackson, MS 39205 (601) 354-7011
Missouri	Dave True Chairman	Missouri Advisory Committee on Arson Prevention Bureau of Alcohol, Tobacco & Firearms 7007 College Boulevard, Suite 410 Overland Park, KS 66211 (9 13) 236-2540

New Jersey	Gerald Maxwell
	Chairman
		New Jersey Advisory Committee on
		  Arson Prevention
		c/o New Jersey Insurance Underwriters
		  Association
		744 Broad Street
		Newark, NJ 07102
		(20l) 622-3838

New York	Francis A. McGarry State Fire Administrator	New York State Office of Fire Prevention and Control c/o New York State Department of State 162 Washington Avenue Albany, NY 12231 (5 18) 474-6746
North Dakota	Steve Devine Chairman	North Dakota Insurance Advisory Council c/o American Family Insurance P.O. Box 966 Fargo, ND 58107 (701) 280-1100
Ohio	David Engleson Executive Secretary	Ohio Blue Ribbon Arson Committee 6230 Busch Boulevard Columbus, OH 43229 (614) 436-4530

STATEWIDE ARSON ORGANIZATIONS (Continued)

State	Contact	Address
Oklahoma	Larry Craig Chairman	Oklahoma Arson Advisory Council c/o Southwest Insurance Information Service 1330 Classen Building Suite 201 Oklahoma City, OK 73106 (405) 236-8612
Oregon	Lt. Rich Hein Commander	Oregon State Police Arson Section 107 Public Service Building Salem, OR 97310 (503) 378-3720
Pennsylvania	Sanford Wilk Coordinator	Pennsylvania Arson Committee c/o State Farm Fire and Casualty Company Willow Grove Service Center 735 Fitzwatertown Road Willow Grove, PA 19090-1385 (215) 784-1814
South Dakota	Leonard Stengal Chairman	South Dakota Anti-Arson, Inc. c/o Stockholm Mutual Insurance Company Box 591 Milbank, SD 57252-059 1 (605) 432-493 1
Texas	Jerry Petree	A Texas Advisory Council on Arson c/o Fire Marshal's Office Midland Fire Department P.O. Box 1152 Midland, TX 79702 (915) 685-7334
Washington	Scott Carpenter Executive Director	Washington Insurance Council 13555 S.E. 36th Street Suite 105 Bellevue, WA 98006 (206) 747-6631

STATEWIDE ARSON ORGANIZATIONS (Continued)

State	Contact	Address
Washington (Cont.)	Steve Smith President	Arson Alarm Foundation 13555 S.E. 36th Street Suite 105 Bellevue, WA 98006 (206) 747-6631
Wisconsin	Leonard H. "Skip" Belstner	Wisconsin Department of Justice Arson Bureau P.O. Box 7857 Madison, WI 53707-7857 (608) 266- 1671
Wyoming	Dave Herring Chairman	Department of Fire Prevention and Electrical Safety Herschler Building 1W Cheyenne, WY 82002 (307) 777-7288

COUNTY ARSON ORGANIZATIONS

County	Contact	Address
California		
Alameda	Donald E. Hawley Inspector	Alameda County Arson Task Force Courthouse, 12th and Fallon Street Oakland, CA 94612 (415) 272-6282
Butte	Norman A. Stump, Sr. Fire Inspector	Butte County Arson Task Force c/o Chico Fire Department 842 Salem Street Chico, CA 95928 (916) 895-4934
Colusa	Randy Dunn Coordinator	Colusa County Arson Task Force c/o Colusa Fire Department 750 Market Street Colusa, CA 95932 (916) 458-7721
Del Norte	Virginia Anthony Detective	Del Norte County Arson Task Force c/o Crescent City Police Department 450 H Street Crescent City, CA 95531 (707) 464-2133
Glenn	James W. Jacobs Deputy Sheriff	Glenn County Arson Task Force c/o Glenn County Sheriffs Office 543 W. Oak Street Willows, CA 95988 (9 16) 934-6431
Humboldt	Dave Parris Coordinator	Humboldt County Arson Task Force Eureka Police Department 604 C Street Eureka, CA 95501 (707) 442-4548
Imperial	Mark Grundman Coordinator	Imperial County Arson Task Force 775 State Street El Centro, CA 92243 (619) 337-4530

COUNTY ARSON ORGANIZATIONS (Continued)

County	Contact	Address
California (Cont.)		
Lassen	Charlie Harrison Team Leader	Lassen County Fire Investigation Task Force P.O. Box 40 Janesville, CA 96114 (916) 253-3737
Marin	Rosemary Bliss Fire Marshal	Marin County Arson Task Force c/o Tiburon Fire Department 1679 Tiburon Boulevard Tiburon, CA 94920 (415) 435-7200
Monterey	R.A. "Ron" Qualls Investigator	Monterey County Fire/Arson Task Force c/o Monterey County Sheriffs Office 1414 Natividad Road Salinas, CA 93906 (408) 755-3773
Napa	Loyde Johnson Battalion Chief	California Department of Forestry Napa County Fire Department 1572 Railroad Avenue St. Helena, CA 94574 (707) 963-3601
Placer	Tony Corado Coordinator	Joint Power Authority Arson Task Force c/o South Placer Fire Department 6900 Eureka Road Roseville, CA 95678 (916) 791-7059
Riverside	Michael Miner	Riverside County Arson Task Force c/o Indio Fire Department 46-990 Jackson Street Indio, CA 92201 (619) 347-0756, Ext: 271

COUNTY ARSON ORGANIZATIONS (Continued)

County	Contact	Address

California (cot.)

County	Contact	Address
San Diego	Eric Stalnaker	Metro Arson Strike Team c/o San Diego Fire Department 1222 First Avenue San Diego, CA 92101 (619) 236-6815
San Joaquin	Steve Thienes Fire Warden	San Joaquin County Arson and Bomb Task Force c/o Bureau of Fire Prevention 222 E. Weber Street Stockton, CA 95202 (209) 468-3380
Sonoma	Doug Williams Coordinator	Sonoma County District Fire Investigation Task Force c/o Bellevue Fire Protection District 207 Todd Road Santa Rosa, CA 95401 (707) 527- 1152
Stanislaus	Russ Richards Fire Warden	Stanislaus County Arson Task Force c/o Stanislaus County Fire Department 929 Oakdale Road Modesto, CA 95355 (209) 525-4657
Tahoe	William Atchley Coordinator	Tahoe Basin Arson Task Force c/o Tahoe City Fire Protection District P.O. Box 1924 Tahoe City, CA 96145 (916) 583-6913

Tuolumne	Ross Jones
	Investigator
		Tuolumne County Fire Investigation Task
		    Force
		c/o Tuolumne County District Attorney's
		    Office
		2 S. Green Street
		Sonora, CA 95370
		(209) 533-5655

COUNTY ARSON ORGANIZATIONS (Continued)

County	Contact	Address
Florida		
Palm Beach	Myron Mansfield District Chief	Palm Beach County Task Force Palm Beach County Fire Rescue Bureau of Fire Investigations 50 S. Military Trail, Suite 101 West Palm Beach, FL 33415 (407) 233-0120
Indiana		
Clark	Ed Shain Director	Clarksville/Clark County Arson Task Force c/o Charlestown Volunteer Fire Department Charlestown, IN 47111 (812) 256-6202
Maryland		
Prince George's	Richard T. Williams Major, Bureau of Fire Investigations	Prince George's County Fire Department 6820 Webster Street Landover Hills, MD 20784 (301) 772-9080
New Jersey		
Atlantic	Edward Hachett Investigator	Atlantic County Prosecutor's Office 19th Avenue and Route 40 P.O. Box 2002 Mays Landing, NJ 08330 (609) 645-7000
Bergen	Lt. Robert Kops Deputy Chief	Bergen County Prosecutor's Office 215 Courthouse Hackensack, NJ 07601 (20 1) 646-2545
Camden	Edward Garrity Sergeant	Camden County Prosecutor's Office Parkade Building 518 Market Street Camden, NJ 08101 (609) 757-8471

County	Contact	Address
New Jersey (Cont.)		
Cape May	Todd Pierce Lieutenant	Cape May County Prosecutor's Office Courthouse Cape May Courthouse, NJ 08210 (609) 465-1135
Essex	Fred Franco Assistant Prosecutor	Essex County Prosecutor's Office Arson Task Force Essex County Courts Building Newark, NJ 07601 (201) 678-0200
Monmouth	Robert Honecker Assistant Prosecutor	Monmouth County Prosecutor's Office Courthouse Freehold, NJ 07728 (201) 431-7160
Union	John Langan Captain	Union County Prosecutor's Office County Administration Building Elizabeth, NJ 07207 (908) 527-4600
New York		
Albany	Terrance K. Ryan Fire Coordinator	Albany County Arson Task Force c/o Public Safety Building Morton Avenue and Broad Street Albany, NY 12202 (5 18) 463-2305
Alleghany	Keith M. Barber Fire Coordinator	Alleghany County Task Force c/o Alleghany County Office Building Belmont, NY 14813 (716) 567-2251
Broome	Thomas J. Vroman Deputy Fire Coordinator	Broome County Arson Task Force Civil Defense Building Upper Front Street Binghamton, NY 13901 (607) 778-2110

County	Contact	Address

New York (Cont.)

County	Contact	Address
Cattaraugus	Stuart B. Lexer Cattaraugus County Fire Coordinator	County Office Building 106% 8th Street Little Valley, NY 14755 (716) 938-9216
Cayuga	Ronald Raymond Cayuga County Fire Coordinator	RD #2, Box 426 Auburn, NY 13021 (3 15) 255-32 11
Chautauqua	Sgt. James Tytka Investigator	Chautauqua County Arson Task Force c/o Sheriffs Department County Jail Mayville, NY 14757 (716) 753-2131
Chemung	Michael S. Smith Fire Coordinator	Chemung County Arson Task Force c/o Justice Building 203-209 William Street Elmira, NY 14901 (607) 737-2097
Chenango	Robert G. Handy Fire Coordinator	Chenango County Arson Task Force c/o Chenango County Communications Center 14 W. Park Place Norwich, NY 13815 (607) 334-5564
Clinton	James King Clinton County Emergency Preparedness Director	Court House Plattsburgh, NY 12901 (518) 565-4791
Columbia	Robert Novak Fire Coordinator	Columbia County Arson Task Force 18 Eichybush Road Kinderhook, NY 12106 (518) 758-7491
Cortland	James E. LeFever Fire Coordinator	Cortland County Arson Task Force 60 Central Avenue, Box 5590 Cortland, NY 13045 (607) 753-5064

County	Contact	Address

New York (Cont.)

County	Contact	Address
Delaware	Nelson Delameter Fire Coordinator	Delaware County Arson Task Force Dept. of Emergency Services Route 1, Box 85D Hamden, NY 13782 (607) 865-7736
Dutchess	Dewitt Sagendorph Fire Coordinator	Dutchess County Arson Task Force c/o Dutchess County Bureau of Fire 348 Creek Road Poughkeepsie, NY 12601 (914) 471-3083
Erie	Vincent Pupo Detective	Erie County Arson Task Force c/o Erie County Sheriffs Office 1 Sheriffs Drive Orchard Park, NY 14127 (716) 662-6150, Ext. 2932
Essex	Raymond C. Thatcher Director of Disaster Preparedness	Essex County Arson Task Force c/o Office of Emergency Services Elizabethtown, NY 12932 (518) 873-6307, Ext. 280
Franklin	Elton W. Cappiello Fire Coordinator	Franklin County Arson Task Force County Office of Emergency Service Bureau of Fire 63 W. Main Street Malone, NY 12932 (5 18) 483-6767
Fulton	Frank A. Bradt Acting Coordinator	Fulton County Fire/Arson Task Force c/o County Complex Building Route 29 and Harrison Street Johnstown, NY 12095 (5 18) 762-3064
Genesee	Lewis P. Schmidt Fire Coordinator	Genesee County Arson Task Force 7690 State Street Road Batavia, NY 14020 (716) 344-0078

County	Contact	Address
New York (Cont.)		
Greene	Ronald Garrison Fire Coordinator	Greene County Arson Task Force Rt. 1, Box 69 Palenville, NY 12463 (518) 943-2515
Hamilton	Clifford Slack Fire Coordinator	Hamilton County Arson Task Force Lake Pleasant, NY 12108 (518) 548-3610
Herkimer	John W. Young Fire Coordinator	Herkimer County Arson Task Force RD 1, Box 306 West Winfield, NY 13487 (315) 867-1212
Jefferson	Richard Madill Fire Coordinator	Jefferson County Arson Task Force County Office Building 175 Arsenal Street Watertown, NY 13601 (3 15) 785-3042
Lewis	William S. Morrow Fire Coordinator	Lewis County Safety Building Stowe Street, P.O. Box 233 Lowville, NY 13367 (3 15) 346-6348
Livingston	David Harter Fire Coordinator	Livingston County Arson Task Force Sheriffs Office 4 Court Street Geneseo, NY 14454 (716) 243-7160
Madison	Joseph DeFrancisco Fire Coordinator	Madison County Arson Task Force c/o Madison County Office Building P.O. Box 577 Wampsville, NY 13163 (3 15) 366-2258

County	Contact	Address
New York (Cont.)		
Monroe	Edward J. Riley Fire Coordinator	Monroe County Arson Task Force c/o Monroe County Bureau of Fire 111 Westfail Road Rochester, NY 14620 (716) 442-6810
Montgomery	Richard Hanson Fire Coordinator	Montgomery County Arson Task Force c/o Montgomery County Office Building Fonda, NY 12068 (518) 829-7818
Nassau	David M. Bartow Fire Marshal	Nassau County Task Force 899 Jerusalem Avenue P.O. Box 128 Uniondale, NY 11553 (5 16) 566-5200
Niagara	Warren J. Rathke Fire Coordinator	Niagara County Arson Task Force 5226 Niagara Street Extension Lockport, NY 14094 (7 16) 439-4071
Oneida	John Paulson Chairman	Oneida County Arson Task Force c/o Utica National Insurance Group New Hartford, NY 13413 (3 15) 734-2000
	Fred VanNamee	Oneida County/Office of Emergency Services 800 Park Avenue Utica, NY 13501 (3 15) 798-5604
Onondaga	Michael S. Waters Fire Coordinator	Onondaga County Arson Task Force c/o Onondaga County Fire Control Center 4694 Central Avenue Syracuse, NY 13215 (3 15) 425-3162

COUNTY ARSON ORGANIZATIONS (Continued)

Counts	Contact	Address

New York (Cont.)

Ontario	Donald Barnes Fire Coordinator	Ontario County Arson Task Force Safety Training Facility 2914 County Road #48 Canandaigua, NY 14424 (716) 396-4310
Orange	David L. Hoffman Fire Coordinator	Orange County Arson Task Force Fire Training Center RD 1, Box 30 New Hampton, NY 10958 (914) 374-7307
Orleans	Richard Clark Fire Coordinator	Orleans County Arson Task Force c/o Civil Defense Headquarters County House Road Albion, NY 14411 (716) 589-4414
Oswego	Wilfred G. Denery Fire Coordinator	Oswego County Arson Task Force 35 E. Cayuga Street Oswego, NY 13126 (3 15) 343-6555
Otsego	Lyle W. Jones, Jr. Fire Coordinator	Otsego County Arson Task Force c/o County Office Building 197 Main Street Cooperstown, NY 13326 (607) 547-5351 (24 Hours) (607) 547-4227 (Office)
Putnam	John Leatter Putnam County Fire Coordinator	Tonetta Lake Park Brewster, NY 10509 (914) 749-1400
Rensselaer	Ivan Wager Fire Coordinator	Rensselaer Arson Task Force Box 72 Berlin, NY 12022 (5 18) 658-2240

COUNTY ARSON ORGANIZATIONS (Continued)

County	Contact	Address
New York (Cont.)		
Rockland	Thomas F. Sullivan Superintendent	Rockland County Arson Task Force c/o Bureau of Criminal Identification and Communications Rockland County Sheriffs Department New City, NY 10956 (914) 638-5456
St. Lawrence	Patrick D. Verschneider Fire Coordinator	St. Lawrence County Arson Task Force Civil Defense Building Court Street Canton, NY 13617 (3 15) 379-2242
Saratoga	Byron J. Baker	Saratoga County Arson Task Force W. High Street Ballston Spa, NY 12020 (518) 885-5381, Ext. 702
Schenectady	Kenneth Posson Fire Coordinator	Schenectady County Arson Task Force 1027 Alheim Drive Schenectady, NY 12303 (5 18) 385-8282
Schoharie	Brian D. Largteau Fire Coordinator	Schoharie County Arson Task Force Schoharie County Office Building Schoharie, NY 12157 (5 18) 295-8344
Schuyler	William S. Randolph Fire Coordinator	Schuyler County Arson Task Force c/o County Jail Building Watkins Glen, NY 14891 (607) 535-4513
Seneca	William Palmer Fire Coordinator	Seneca County Arson Task Force 7123 Orchard Street Ovid, NY 14521 (607) 869-5417

County	Contact	Address
New York (Cont.)		
Steuben	Donald Merring Fire Coordinator	Steuben County Arson Task Force County Office Building 3 Pulteney Square Bath, NY 14810 (607) 776-9631, Ext. 2394
Suffolk	Al Jardin Chief Fire Marshal	Suffolk County Arson Task Force c/o Department of Fire, Rescue, and Emergency Services Yaphank Avenue, Box 95 Yaphank, NY 11980 (5 16) 345-6350
Sullivan	Harold Kronenburg Fire Coordinator	Sullivan County Arson Task Force Bureau of Fire County Airport, P.O. Box 109 White Lake, NY 12786 (914) 583-7127
Tioga	Kenneth G. Wolff Fire Coordinator	Tioga County Arson Task Force 26 Talcott Street Owego, NY 13827 (607) 687-1235
Tompkins	John L. Miller Fire Coordinator	Tompkins County Airport Ithaca, NY 14850 (607) 257-3888
Ulster	Duncan Wilson Fire Coordinator	Ulster County Arson Task Force Box 158 Bearsville, NY 12401 (914) 331-1216
Warren	Marvin F. Lemery Fire Coordinator	Warren County Arson Task Force c/o Municipal Center Fire Prevention and Control Department Lake George, NY 12845 (518) 761-6497

COUNTY ARSON ORGANIZATIONS (Continued)

County	Contact	Address

New York (Cont.)

County	Contact	Address
Washington	Robert Potter Fire Coordinator	Washington County Arson Task Force Gillis Hill Road Salem, NY 12865 (5 18) 854-3780
Wayne	Richard Bond Fire Coordinator	Wayne County Arson Task Force 7336 Route #31 Lyons, NY 14489 (3 15) 946-5641
Westchester	Walter L. Groden	Westchester County Arson Task Force c/o Westchester County Fire Training Center Dana Road Valhalla, NY 10595 (914) 592-8145
Wyoming	Jack Fisher Fire Coordinator	Wyoming County Arson Task Force c/o Wyoming County Courthouse Public Safety Building 143 N. Main Street Warsaw, NY 14569 (7 16) 786-8867
Yates	Glen R. Miller Fire Coordinator	Yates County Arson Task Force Public Safety Building 227 Main Street Penn Yan, NY 14527 (3 15) 536-3000

North Carolina

County	Contact	Address
Buncombe	Harley Shuford Director	Asheville-Buncombe Arson Task Force P.O. Box 7148 Asheville, NC 28802 (704) 259-5648

24

County	Contact	Address
North Carolina (Cont.)		
Burke	Clinton Patton Director, Emergency Services	Burke County Office of Emergency Management P.O. Box 219 Morgantown, NC 28655 (704) 437-3039
	Mike Long Investigator	Burke County Arson Task Force Burke County Office of Emergency Services P.O. Box 219 Morgantown, NC 28655 (704) 437-3039
	Mark Pitts Investigator	Burke County Arson Task Force Burke County Office of Emergency Services P.O. Box 219 Morgantown, NC 28655 (704) 437-3039
	H.K. Stewart Investigator	Burke County Arson Task Force Burke County Office of Emergency Services P.O. Box 219 Morgantown, NC 28655 (704) 437-3039
Caldwell	Dale Coffey Investigator	Caldwell Fire Arson Task Force Caldwell County Fire Department P.O. Box 2200 Lenoir, NC 28645 (704) 757-1277
Catawba	Rupert H. Little Investigator/Fire Marshal	Catawba County Fire Marshal P.O. Box 389 Newton, NC 28658 (704) 465-8230

County	Contact	Address

North Carolina (Cont.)

Cleveland	R.L. Lovelace Cleveland County Fire Marshal	Cleveland County Arson Task Force Shelby, NC 28150 (704) 484-4841
Macon	Farrell Jamison Chief Investigator	Macon County Investigation Task Force P.O. Box 177 Franklin, NC 28735 (704) 524-8444
Watauga	Charles Collins Fire Marshal	Watauga County Fire Marshal 301 W. Queen Street Boone, NC 28607 (704) 264-4235

Oregon

Central Oregon	Jim Miller	Central Oregon Fire Investigation Team c/o Oregon State Police 63319 Highway 20-W Bend, OR 97701 (503) 388-6303
Clackamas	Steve Sasser Sergeant	Clackamas County Arson Strike Force c/o Oregon State Police P.O. Box 66009 Portland, OR 97266 (503) 238-8434
Klamath	Jim Miller Detective	Klamath Area Fire Investigation Strike Team c/o Oregon State Police 63319 Highway 20-W Bend, OR 97701 (503) 388-6303

COUNTY ARSON ORGANIZATIONS (Continued)

County	Contact	Address
Oregon (cont.)		
Marion/Polk	Larry Kirk	Marion/Polk Fire Investigation Team 300 Cordon Road, N.E. Salem, OR 97301 (503) 588-6508
Mid-Columbia	Fred Hawkins Detective	Mid-Columbia Arson Strike Force c/o Oregon State Police 3313 North East Frontage Road The Dalles, OR 97058 (503) 296-2161 (24 Hours) (503) 296-9646 (8 AM - 5 PM)
Yamhill	Steve Sasser Sergeant	Yamhill County Arson Strike Force c/o Oregon State Police P.O. Box 66009 Portland, OR 97266 (503) 238-8434
Texas		
Tarrant	Ron Harthorne Secretary	Tarrant County Arson Task Force P.O. Box 157 Bedford, TX 76095 (817) 952-2140
Washington		
Bellingham/Whatcom	Robert Neale Fire Marshal	Bellingham/Whatcom County Arson Task Force 210 Lottie Street 1800 Broadway Bellingham, WA 98225 (206) 6766832
Chelan/Douglas	Glenn Tibbs Fire Marshal	Chelan/Douglas County Task Force on Arson and Fire Investigation c/o Wenatchee Fire Department 136 S. Chelan Avenue Wenatchee, WA 98801 (509) 6626125

Counts	Contact	Address
Washington (Cont.)		
King	Richard Hargett Assistant Fire Marshal	King County Arson Task Force 3600 136 Place, S.E. Bellevue, WA 98006-1400 (206) 296-6670
Pierce	David Kreg Detective	Tacoma Police Department Pierce County Public Safety Building 930 Tacoma Avenue, S. Tacoma, WA 98402 (206) 591-5815
Wisconsin		
Brown	Russ Hawley Investigator	Brown County Fire Investigation Task Force c/o Brown County Law Enforcement Center 300 E. Walnut Green Bay, WI 54301
Calumet	Oscar Bielke, Jr. Coordinator	Calumet County Arson Task Force 206 Court Street Chilton, WI 53014 (414) 849-2335
Central Wisconsin	Robert Pound Coordinator	Central Wisconsin Arson Task Force 1000 Campus Drive Wausau, WI 54401 (715) 675-3331
Dane	Willis Hartwig Detective	Dane County Arson Task Force City-County Building Room GR-20 Madison, WI 53709 (608) 266-4930
Kenosha	Paul Guilbert, Jr. Chairman	Kenosha County Arson Task Force P.O. Box 89 Pleasant Prairie, WI 53158 (414) 694-4066

County	Contact	Address
Wisconsin (Cont.)		
Manitowoc	Arik Hansen Coordinator	Manitowoc County Fire Investigation Unit 7232 Manitou Drive Two Rivers, WI 54241 (4 14) 682-4776
Gutagamie	Gene Sipple Coordinator	Outagamie County Arson Task Force 320 S. Walnut Street Appleton, WI 54911 (414) 832-5605
Portage	David Page Coordinator	Portage County Fire Department Task Force 5255 Jordan Road Stevens Point, WI 54481 (715) 346-7012
Racine	Lawrence "Bud" Eastman Coordinator	Racine County Arson Task Force 201 First Street Racine, WI 53403 (414) 637-7681
Walworth	Detective David Fladten	Walworth County Sheriffs Dept. County Courthouse Elkhorn, WI 53121 (414) 741-4400
Wood	Mark Gosh Coordinator	Wood County Arson Task Force 400 Market Street P.O. Box 884 Wisconsin Rapids, WI 54494 (715) 421-8700

METROPOLITAN ARSON ORGANIZATIONS

City	Contact	Address
Alabama		
Florence	Charles Cochran Fire Marshal	Arson Strike Force 402 S. Wood Avenue Florence, AL 35630 (205) 760-6475
Huntsville	R.W. Schrimsher Fire Marshal	Arson Strike Force P.O. Box 308 Huntsville, AL 35804 (205) 532-7474
Arkansas		
Little Rock	Jack Ballard Chief, Fire Prevention/ Training	Little Rock Fire Department 1000 W. Seventh Street Little Rock, AR 72201-3995 (501) 371-4796
Connecticut		
Bridgeport	David Rentz Deputy Fire Marshal	Bridgeport Arson Task Force 30 Congress Street Bridgeport, CT 06604 (203) 576-8269
Enfield	Paul J. Skowron Chairman	Enfield Arson Task Force 820 Enfield Street Enfield, CT 06082 (203) 745-1671
New Haven	John E. Smith Fire Chief	New Haven Arson Task Force 952 Grand Avenue New Haven, CT 06510 (203) 787-6300
Waterbury	Edward D. Bergin Mayor	Waterbury Arson Task Force Waterbury City Hall 236 Grand Street Waterbury, CT 06702 (203) 597-3452

METROPOLITAN ARSON ORGANIZATIONS (Continued)

City	Contact	Address
Florida		
Cleat-water	James E. Goodloe Fire Marshal	Countywide Arson Task Force 610 Franklin Street P.O. Box 4748 Clearwater, FL 34618 (813) 462-6300
Jacksonville	George H. Smith Fire Marshal	Mayor's Arson Task Force 1931 E. Beaver Street Jacksonville, FL 32202 (904) 630-0969
Georgia		
Atlanta	J.A. Haynie Chief of Investigations	Atlanta Fire Department Arson Investigations 46 Courtland Street, S.E. Atlanta, GA 30335 (404) 658-6904
Illinois		
Chicago	William C. Alletto Deputy Fire Commissioner	Chicago Fire Department Bureau of Support Services Administrative Center Office of Fire Investigations 3015 W. 31st Street Chicago, IL 60623 (3 12) 747-3982 (312) 747-6594 (Fax)
Harvey	David Riddle Deputy Chief	Harvey Arson Task Force 15600 Center Avenue Harvey, IL 60426 (708) 331-7720

City	Contact	Address
Iowa		
Des Moines	Kenneth Danley Fire Marshal	Des Moines Arson Task Force 900 Mulberry Street Des Moines, IA 50309-3614 (515) 283-4198
Louisiana		
Baton Rouge	Glenn J. Sevier Chief Fire Investigator	Baton Rouge Fire Department Arson Division 805 St. Louis Baton Rouge, LA 70821 (504) 389-4664
New Orleans	Walter Dupeire Coordinator	New Orleans Arson Task Force 317 Decatur Street New Orleans, LA 70127 (504)565-7827
New York		
Albany	James Larsen Chief	Albany Fire Department 26 Broad Street Albany, NY 12202 (518) 447-78/77, 78, 79
	Salvatore Prividera Captain	Albany Arson Task Force c/o Albany Fire Department Public Safety Building Albany, NY 12202 (518)462-8679
Albion	Donald Hinman Acting Chief	Albion Arson Task Force c/o Albion Police Department 301 Washington Street Albion, NY 14411 (716) 589-5626

City	Contact	Address
New York (Cont.)		
Amsterdam	Jim LaConte Battalion Chief	Amsterdam Arson Task Force c/o Amsterdam Fire Department Public Safety Building Amsterdam, NY 12010 (518) 843-1312
	James Scheckton Chief	Amsterdam Fire Department Public Safety Building Amsterdam, NY 12010 (518) 843-1313
Auburn	Frank A. Calarco Chief	Auburn Arson Task Force c/o Auburn Fire Department 23 Market Street Auburn, NY 13021 (3 15) 253-32 11
Batavia	Keith W. Hunt Chief	Batavia Arson Task Force c/o Batavia Fire Department Fire Headquarters 18 Evans Street Batavia, NY 14020 (716) 343-8180, Ext. 41
Binghamton	James Malane Chief	Binghamton Fire Department Government Administration Building 38 Hawley Street Binghamton, NY 13901 (607) 772-7016
	Gerald Tita Fire Marshal	Binghamton Arson Task Force c/o Binghamton Fire Department Government Administrative Building 38 Hawley Street Binghamton, NY 13901 (607) 772-70 16

METROPOLITAN ARSON ORGANIZATIONS (Continued)

City	Contact	Address

New York (Cont.)

City	Contact	Address
Buffalo	Paul Shanks Chief	Buffalo Arson Task Force c/o Buffalo Fire Department 195 Court Street Buffalo, NY 14202 (716) 855-5333
Canandaigua	James Farrell Chief	Canandaigua Arson Task Force c/o Canandaigua Fire Department 13 Niagara Street Canandaigua, NY 14424 (716) 394-2111
Cohoes	Andrew J. Gisondi Fire Chief	Cohoes Arson Task Force c/o City Hall Cohoes, NY 12047 (518) 237-5011
Corning	William Flohr Chief	Corning Arson Task Force c/o Corning Fire Department 1 Corning Boulevard Corning, NY 14830 (607) 962-3 151
Cortland	Dennis M. Baron Chief	Cortland Arson Task Force c/o Cortland Fire Department 21 Court Street Cortland, NY 13045 (607) 756-56 13
East Greenbush	Chris Lavin Chief of Police	East Greenbush Arson Task Force c/o East Greenbush Police Department Detective Division 82 Columbia Turnpike Rensselaer, NY 12144 (5 18) 479-2525
Elmira	Donald H. Harrison Chief	Elmira Arson Task Force c/o Elmira Fire Department 101-109 W. Second Street Elmira, NY 14901 (607) 737-5700

METROPOLITAN ARSON ORGANIZATIONS (Continued)

City	Contact	Address

New York (Cont.)

Endicott	Paul Ripic Coordinator	Endicott Arson Task Force 224 Madison Avenue Endicott, NY 13760 (607) 785-3385
Geneva	Ralph E. DeBolt Chief	Geneva Arson Task Force c/o Geneva Fire Department City Hall Castle Street Geneva, NY 14456 (315) 789-2121
Glens Falls	Elwood Greene Lieutenant	Glens Falls Fire Department 134 Ridge Street Glens Falls, NY 12801 (518) 792-1133
Gloversville	Mark Pollak Fire Marshal	Gloversville Arson Task Force c/o Gloversville Fire Department Frontage Road Gloversville, NY 12078 (518) 725-3122
	Mike Shafer Acting Chief	Gloversville Fire Department Frontage Road Gloversville, NY 12078 (518) 725-3122
Greenburg	Robert Mauro Chief	Greenburg Arson Task Force c/o Fairview Fire Department 19 Rosemont Boulevard White Plains, NY 10607 (914) 949-2828
Homell	Vincent L. Kelly Chief	Hornell Arson Task Force c/o Hornell Fire Department 110 Broadway Hornell, NY 14623 (607) 324-2100

City	Contact	Address

New York (Cont.)

Hudson	Edward A. Eisley Chief	Hudson Arson Task Force c/o Hudson Police Department 427 Warren Street Hudson, NY 12534 (518) 828-3388
Ithaca	Edward Olmstead Chief	Ithaca Arson Task Force c/o Ithaca Fire Department 310 W. Green Street Ithaca, NY 14850 (607) 272-1234
Jamestown	Charles Hadjuk Chief	Jamestown Fire Department E. 2nd Street Jamestown, NY 14701 (716) 483-7958
Johnson City	George Maney Chief	Johnson City Arson Task Force c/o Johnson City Fire Department 270 Floral Avenue Johnson City, NY 13790 (607) 729-9512
Johnstown	Edward Heberer Chief	Johnstown Arson Task Force 244 N. Perry Street Johnstown, NY 12095 (518) 762-4438
Kingston	John Reinhardt Chief	Kingston Fire Department 19 E. O'Reilly Street Kingston, NY 12401 (914) 331-1216
	Chris Rea Lieutenant	Kingston Arson Task Force c/o Kingston Fire Department 19 E. O'Reilly Street Kingston, NY 12401 (914) 331-1216

City	Contact	Address
New York (Cont.)		
LeRoy	Russell Lathan Chief	LeRoy Arson Task Force c/o LeRoy Fire Department 3 W. Main Street LeRoy, NY 14482 (7 16) 768-2527
Lockport	Tom Darroch Chief	Lockport Fire Department One Locks Plaza Lockport, NY 14094 (716) 439-6611
Mount Vernon	Henry Campbell Chief	Mount Vernon Fire Department 470 E. Lincoln Avenue Mount Vernon, NY 10652 (9 14) 688-9002
Newburgh	James Barry Chief	Newburgh Fire Department 55 Broadway Newburgh, NY 12550 (9 14) 565-4995
	Robert Paden Deputy Chief	Newburgh Arson Task Force c/o Newburgh Fire Department 22 Grand Street Newburgh, NY 12550 (9 14) 565-4942
New Rochelle	Anthony Fillat Lieutenant	New Rochelle Arson Task Force c/o New Rochelle Fire Department 90 Beaufort Place New Rochelle, NY 10801 (9 14) 654-22 12
New York	Barbara Shulman Coordinator	New York City Arson Strike Force 250 Livingston Street Brooklyn, NY 10007 (212) 566-7263

METROPOLITAN ARSON ORGANIZATIONS (Continued)

City	Contact	Address
New York (Cont.)		
New York (Cont.)	John J. Stickevers Chief Fire Marshal	Bureau of Fire Investigation New York City Fire Department 250 Livingston Street Brooklyn, NY 1120l-5884 (212) 403-1434
Niagara Falls	John L. Gabriele Chief	Niagara Falls Arson Task Force P.O. Box 69 520 Hyde Park Boulevard Niagara Falls, NY 14302-0069 (7 16) 286-4725
North Tonawanda	Chief David Rogge	495 Zimmerman Street North Tonawanda, NY 14120 (716) 693-2341
Ogdensburg	Richard Fournier Chief	Ogdensburg Arson Task Force c/o Ogdensburg Fire Department Central Fire Station 701 Ford Street Ogdensburg, NY 13669 (3 15) 393-232 1
Olean	Chief John Gibbons or Captain David Grosse	Olean Fire Department 542 N. Union Street Olean, NY 14760 (716) 372-3212 (7 16) 375-5686
Oneida	Erwin Smith Chief	Oneida Arson Task Force c/o Oneida Fire Department 109 N. Main Street Oneida, NY 13421 (315) 363-1910
Oswego	James Borden Chief	City of Oswego Fire Investigation Team c/o Oswego Fire Department 35 E. Cayuga Street Oswego, NY 13126 (315) 343-2161

City	Contact	Address

New York (Cont.)

Peekskill	Robert L. Ferris Chief	Peekskill Fire Department 828 Main Street Peekskill, NY 10566 (9 14) 737-2760
	Detective David A. Levine Arson Investigator	2 Nelson Avenue Peekskill, NY 10566 (9 14) 737-8000
Plattsburgh	James H. Squires Chief	Plattsburgh Arson Task Force c/o Plattsburgh Fire Department 65 Cornelia Street Plattsburgh, NY 12901 (518) 561-5965
Poughkeepsie	Dennis C. McComb Chief	Poughkeepsie Arson Task Force c/o Poughkeepsie Fire Department City Hall Complex Market Street Poughkeepsie, NY 12601 (914) 431-8330
Rensselaer	J. Michael Doring Chief	Rensselaer Arson Task Force c/o Rensselaer Fire Department Station #2 505 Broadway Rensselaer, NY 12144 (5 18) 465-3243
Rochester	Charles D. Ippolito	Rochester Fire Department 306 Public Safety Building 150 S. Plymouth Avenue Rochester, NY 14614 (716) 428-7485
	William Kelly Captain	Rochester Arson Task Force c/o Rochester Fire Department 306 Public Safety Building 150 S. Plymouth Avenue Rochester, NY 14614 (716) 428-7485

City	Contact	Address

New York (Cont.)

City	Contact	Address
Rome	Ronald Swinney Chief	Rome Arson Task Force c/o Rome Fire Department City Hall Rome, NY 13440 (315) 336-1237
Salamanca	John M. McClune Chief	Salamanca Arson Task Force c/o Salamanca Fire Department 225 Wildwood Avenue Salamanca, NY 14779 (716) 945-33 11
Saratoga Springs	Vincent Camarro Chief	Saratoga Springs Fire Department 60 Lake Avenue Saratoga Springs, NY 12866 (518) 587-3599
	Paul Varley Assistant Chief	Saratoga Springs Arson Task Force c/o Saratoga Springs Fire Department 60 Lake Avenue Saratoga Springs, NY 12866 (518) 587-3599
Scarsdale	Walter F. Felice Chief	Scarsdale Arson Task Force c/o Scarsdale Fire Department Village Hall Scarsdale, NY 10583 (914) 723-2514
Schenectady	Thomas Varno Chief	Schenectady Arson Task Force c/o Schenectady Fire Department Veeder Avenue Schenectady, NY 12307 (518) 382-5140
Schodack	Sgt. Barry Secor	Schodack Police Department 1777 Columbia Turnpike Castleton-on-Hudson, NY 12144 (5 18) 477-7611

City	Contact	Address
New York (Cont.)		
South Hampton	Kenneth H. Jones Chief Fire Marshal	116 Hampton Road South Hampton, NY 11968 (516) 288-0201
Syracuse	J. Michael Hewitt Lieutenant	Syracuse Arson Task Force 511 S. State Street, Room 609 Syracuse, NY 13202 (3 15) 473-3296
	Donald Moriarty Chief	Syracuse Fire Department 511 S. State Street Syracuse, NY 13202 (3 15) 473-5525
Tonawanda	Assistant Chief Curtis Mesler Coordinator	Tonawanda Arson Task Force 44 William Street Tonawanda, NY 14150 (7 16) 692-8400
	Thomas D. Miller Chief	Tonawanda Fire Department 44 William Street Tonawanda, NY 14150 (7 16) 692-8400
Troy	Edward Shultz Chief	Troy Arson Task Force c/o Troy Fire Department 55 State Street Troy, NY 12180 (5 18) 270-4442
Utica	Anthony Cuccinotta Coordinator	Utica Arson Task Force c/o Utica Police Department Oriskany Street, W. Utica, NY 13502 (315) 735-3301

City	Contact	Address
New York (Cont.)		
Watertown	Ronald Chisamore Chief	Watertown Fire Department 217 Arsenal Street Watertown, NY 13601 (315) 782-6060
	Edward J. Lachenauer Captain, Senior Investigator	Watertown Arson Task Force c/o Watertown Fire Department 224 Massey Street Watertown, NY 13601 (3 15) 782-6060
White Plains	John Cullen Chief	White Plains Arson Task Force c/o White Plains Fire Department 219 Mamaroneck Avenue White Plains, NY 10605 (9 14) 949-6302
Yonkers	Donald Starkey Lieutenant	Yonkers Arson Task Force c/o Yonkers Fire Department Arson Investigation Unit 7 School Street Yonkers, NY 10701 (9 14) 377-7500
North Carolina		
Charlotte	David Lowery Chief Fire Investigator	Charlotte Fire Investigation Task Force 1215 South Boulevard Charlotte, NC 28203 (704) 336-3970
Hickory	Tom R. Bradshaw Fire Marshal	Hickory Arson Task Force 19 Second Street Drive, N.E. Hickory, NC 28601 (704) 323-7522

City	Contact	Address
North Carolina (Cont.)		
Morehead City	Dave Marshal SBI Special Agent	Morehead City - Atlantic Beach Joint Task Force P.O. Box 10 Atlantic Beach, NC 28512 (919) 726-7361
New Bern	Captain Henry Watson Chief Investigator	New Bern Fire Investigation Unit 420 Broad Street New Bern, NC 28560 (9 19) 636-4020
Wilmington	E.E. Benton Chief	Wilmington Fire Investigation Task Force 20 S. Fourth Street Wilmington, NC 28401 (919) 341-7846
North Dakota		
Fargo	Dan Freeman Captain	Fargo Arson Task Force 637 N.P. Avenue Fargo, ND 58102 (701) 241-1540
Oregon		
Klamath Falls	Lyle Mahan Fire Chief	Klamath Area Arson Task Force Klamath Falls Fire Department 4042 Balsam Drive Klamath Falls, OR 97601 (503) 884- 1670
Bend	Ron Pugh Special Agent	Central Oregon Fire Investigation Team Deschutes National Forest 1645 Highway 20E Bend, OR 97701 (503) 383-5510

City	Contact	Address
Oregon (cont.)		
Ontario	Allan Higginbotbam Deputy Fire Marshal	Snake River Fire Investigation Team 325 Goodfellow Street Ontario, OR 97914 (503) 889-6469
Rhode Island		
Coventry	Captain Ernest Quaglieri Chief of Detectives	Coventry Arson Task Force 1075 Main Street Coventry, RI 02816 (401) 826-1100
Tiverton	Peter Lamb Chief	Tiverton Arson Task Force Tiverton Fire Department 1449 Main Road Tiverton, RI 02878 (401) 625-6740
Tennessee		
Memphis	Bill Wallace Captain	Metro Arson Task Force Memphis Fire Department 2201 Lamar Memphis, TN 38114 (901) 320-5465
Wisconsin		
Menasha	Joseph W. Lingnofski Fire/Arson Investigator	Menasha Police Department 430 First Street Menasha, WI 54952 (414) 729-5144

THE NATIONAL FIRE INCIDENT REPORTING SYSTEM (NFIRS)

NFIRS is the largest fire incident specific data collection effort in the United States, It is operated by the U.S. Fire Administration (USFA/FEMA) in conjunction with the National Fire Information Council (NFIC) and the voluntarily participating States and local fire departments. As currently configured, this system provides a core of standardized fire incident, firefighter, and civilian casualty data unavailable from any other source, for the quantification and analysis of fire loss experience at the local, State, and Federal level.

The 40 participating States currently represent approximately 12,000 fire departments generating in excess of 1,000,000 incident records annually. The system has enjoyed expanding participation since its inception in 1976 when six States piloted the initial NFIRS prototype. NFIRS now offers participants both mainframe and microcomputer applications for data collection and analysis, thus allowing each level of participant to utilize the data collected by their reporting jurisdiction(s).

The obvious strength of this system is the local fire departments' commitment of standardized fire incident data collection, which allows comparison of their fire experience to the experience of others. NFIRS provides the common ground for the detailed examination of the Nation's fire problem, through a cooperative effort by fire departments and State governments which could be accomplished by no other known means.

NATIONAL FIRE INCIDENT REPORTING SYSTEM

Federal	Contact	Address
FEMA	Harold Collins Computer Program Analyst	Federal Emergency Management Agency 5321 Riggs Road Gaithersburg, MD 20879 (301) 926-5376
FEMA/USFA	Ken Kuntz Fire Studies Specialist	Federal Emergency Management Agency/U.S Fire Administration 16825 S. Seton Avenue Emmitsburg, MD 21727 (301) 447-1271

NATIONAL FIRE INCIDENT REPORTING SYSTEM

State	Contact	Address
Alaska	Kenneth Lea	Alaska Fire Marshal Office P.O. Box 111200 Juneau, AK 99811 (907) 465-4331
Arkansas	Sheldon Richardson Director	Arkansas Fire Academy SAU-Tech P.O. Box 3499 East Camden, AR 71701 (501) 574-1521
Connecticut	Wayne H. Maheu Director	Training and Fire Analysis Division of Fire and Building Safety Department of Public Safety 294 Colony Street Meriden, CT 06450 (203) 238-6257
Delaware	Daniel R. Kiley	Delaware State Fire Marshal RT #2, Box 166A Dover, DE 19901 (302) 739-5665
District of Columbia	Lt. Donald Wood	District of Columbia Fire and Emergency Medical Service 1923 Vermont Avenue, N.W. Washington, DC 20001 (202) 673-3320
Florida	Patricia A. Gibney	Insurance Department Fire Marshal Division 652 Fletcher Tallahassee, FL 32399-0300 (904) 922-3172, Ext. 3630
Georgia	A.D. Bell	Georgia State Fire Marshal #2 Martin Luther King Jr. Drive West Tower, Room 620 Atlanta, GA 30334 (404) 656-2064

NATIONAL FIRE INCIDENT REPORTING SYSTEM (Continued)

State	Contact	Address
Hawaii	August Range	Honolulu Fire Department 3375 Koapaka Street Suite H425 Honolulu, HI 96819 (808) 831-7748
Idaho	Lee Bright	Idaho State Fire Marshal 700 W. State Street Boise, ID 83720 (208) 334-4370
Illinois	Barbara J. Petrilli	Illinois State Fire Marshal 1035 Stevenson Drive Springfield, IL 62703 (217) 785-1016
Indiana	Mike Newsom	Indiana Department of Fire Prevention 402 W. Washington Street Room 241E Indianapolis, IN 46204 (3 17) 232-6236
Iowa	Karen Shipley	Iowa State Fire Marshal E. 9th and Grand, Wallace Building Des Moines, IA 50319 (515) 281-7003
Kansas	Karl W. McNorton	Kansas State Fire Marshal 700 Southwest Jackson, ST600 Topeka, KS 66603-3714 (913) 296-3401
Kentucky	Rodney Raby	Kentucky State Fire Marshal 127 S. Building; US 127 South Frankfort, KY 40601 (502) 564-3626
Louisiana	Michael Cammarosanno Administrative Director	Louisiana State Fire Marshal 5150 Florida Boulevard Baton Rouge, LA 70806 (504) 925-4911

NATIONAL FIRE INCIDENT REPORTING SYSTEM (Continued)

State	Contact	Address
Maine	Dermis Lundstedt	Maine State Fire Marshal 317 State Street, Station #52 Augusta, ME 04333-0052 (207) 287-3473
Maryland	Rocco J. Gabriele	Maryland State Fire Marshal 106 Old Court Road, Suite 300 Pikesville, MD 21208-4016 (401) 764-4324 (800) 525-3124
Massachusetts	Marty Ahrens Research Analyst	Office of the State Fire Marshal 1010 Commonwealth Avenue Boston, MA 02215 (617) 566-4500, Ext. 257
Michigan	Robert Jensen	Michigan State Fire Marshal 7150 Harris Drive East Lansing, MI 48913 (5 17) 322-5460
Minnesota	Thomas R. Brace	Minnesota State Fire Marshal 285 Bigelow Building 450 N. Syndicate Street Saint Paul, MN 55104 (612) 643-3080
Montana	Anita Varone	Fire Prevention and Investigation Bureau Scott Hart Building P.O. Box 201417 Helena, MT 59620-1417 (406) 444-2050
Nebraska	Lori Loyd	Nebraska State Fire Marshal 246 S. 14th Street Lincoln, NE 68509 (402) 471-2027
New Hampshire	Donald P. Bliss	New Hampshire State Fire Marshal 10 Hazen Drive Concord, NH 03305 (603) 271-3294

NATIONAL FIRE INCIDENT REPORTING SYSTEM (Continued)

State	Contact	Address
New Jersey	John McQuade	New Jersey State Fire Marshal 333 North Hampton Drive Willingboro, NJ 08046 (609) 871-2867
New York	Jim Brizzell	New York Office of Fire Prevention and Control 162 Washington Avenue Albany, NY 12231 (5 18) 474-6746
North Carolina	Tim Bradley	North Carolina Fire Commission P.O. Box 26387 Raleigh, NC 27611 (919) 733-5435
Ohio	Charles G. McGrath Chief, Arson Bureau	Ohio State Fire Marshal 8895 E. Main Street Reynoldsburg, OH 43068 (614) 752-8299
Oregon	Everett Hall	Oregon State Fire Marshal 4760 Portland Road, N.E. Salem, OR 97305-1760 (503) 378-3473
Rhode Island	Donald Byrne	Rhode Island State Fire Marshal 272 W. Exchange Street Providence, RI 02903 (401) 277-2335
South Carolina	Bill Graham	South Carolina Law Enforcement Division Broad River Road Columbia, SC 29210 (803) 737-3166
South Dakota	Helen King	South Dakota State Fire Marshal's Office 118 W. Capitol Pierre, SD 57501 (605) 773-4311

State	Contact	Address
Tennessee	W. Robert Frost Assistant Commissioner	500 James Robertson Parkway Third Floor Nashville, TN 37243-0577 (615) 741-2981
Texas	Jean Mitchell	Texas Commission on Fire Protection P.O. Box 2286 Austin, TX 78758 (512) 873-1784
Utah	Janet Cherron	Utah State Fire Marshal 4501 S. 2700 West Street Salt Lake City, UT 84119 (801) 965-4353
Vermont	Lt. Clayton Perkins	Department of Public Safety Office of the Fire Marshal 103 S. Main Street Waterbury, VT 05671 (802) 244-8781
Virginia	Marion A. Long	Department of Fire Programs 2807 Parham Road Suite 200 Richmond, VA 23294 (804) 527-4236
Washington	R. Kent Dewitt	Washington State Fire Marshal Fire Protection Services P.O. Box 48350 Olympia, WA 98504-8350 (206) 493-2655
West Virginia	Paul Gill	West Virginia State Fire Marshal Fire, Arson, and Explosive Division 2100 Washington Street, E. Charleston, WV 25305 (304) 558-2 191 (304) 558-2537 (Fax)

NATIONAL FIRE INCIDENT REPORTING SYSTEM (Continued)

State	Contact	Address
Wisconsin	Karen Johnson	201 E. Washington Avenue P.O. Box 7969 Madison, WI 53707-7969 (608) 267-5264
Wyoming	Richard DuBay	Department of Fire Prevention and Electrical Safety Herschler Building 1W Cheyenne, WY 82002 (307) 777-6683

NATIONAL FIRE INCIDENT REPORTING SYSTEM

Metro	Contact	Address
Arizona		
Phoenix	Stephen Kreis	Phoenix Fire Department 620 W. Washington Street Room 465 Phoenix, AZ 85003 (602) 256-3394
California		
El Monte	Capt. William Franklin	Los Angeles County Fire Department 5110 Peck Road El Monte, CA 91732 (8 18) 450-7460
Los Angeles	Dal Howard Assistant Fire Marshal	Los Angeles City Fire Department 200 N. Main Street Los Angeles, CA 90012 (2 13) 485-5980
Orange	Bruce M. Hunt	Orange County Fire Department 180 S. Water Street Orange, CA 92666 (714) 744-0511
Sacramento	David LeMay	California Department of Forestry P.O. Box 944246 Sacramento, CA 94244 (9 16) 653-9747
San Diego	Ron Darrah	San Diego Fire Department 525 B Street, Suite 805 San Diego, CA 92101 (619) 533-4300
San Francisco	Gary Tortes	San Francisco Fire Department 260 Golden Gate Avenue San Francisco, CA 94102 (415) 861-8000, Ext. 0272 (415) 431-9655 (Fax)

Metro	Contact	Address

California (Cont.)

San Jose · Donald Kelley · San Jose Fire Department
4 N. Second Street, Suite 1100
San Jose, CA 95110
(408) 277-4444

Colorado

Aurora · Leslie W. Schlitt · Aurora Fire Department
1470 S. Havanna Street, Suite 420
Aurora, CO 80012
(303) 695-7118

Denver · Lt. J.V. Sarconi · Denver Fire Department
Arson Bureau
745 W. Colfax Avenue
Denver, CO 80204
(303) 575-3435

Illinois

Chicago · Frank Barbaro · Chicago Fire Department
510 N. Peshtigo Court
Chicago, IL 60611
(3 12) 744-8654

Kentucky

Louisville · Donald Cummins · Louisville Fire Department
1135 W. Jefferson Street
Louisville, KY 40203
(502) 625-3711

Massachusetts

Boston · Stephen Morash · Boston Fire Department
115 Southampton Street
Boston, MA 02118
(617) 725-3483

Metro	Contact	Address
Maryland		
Landover	Gene Berry	Prince George's County Fire Department Bureau of Support Services 7911 Anchor Street, Room 222 Landover, MD 20785 (301) 499-8130
Michigan		
Detroit	J. Richard Milliner Fire Marshal	Detroit Fire Department 250 W. Lamed Detroit, MI 48226 (3 13) 596-2926
New York		
Rochester	Edward Reiley	Monroe County Fire Department 111 Westfall Road Rochester, NY 14620 (716) 442-6810
Ohio		
Cleveland	Alvin Cipra	Cleveland Fire Department 1645 Superior Avenue Cleveland, OH 44114 (216) 241-2524
Columbus	Timothy Carty	Columbus Fire Department 300 N. Fourth Street Columbus, OH 43215 (614) 645-7641
Texas		
Dallas	William D. Morrison	Dallas Fire Department 1500 Marilla Street, Room L1CS Dallas, TX 75201 (2 14) 670-4565

Metro	Contact	Address

Texas (Cont.)

Houston	Donald Clark	Houston Fire Department 410 Bagby Street Houston, TX 77002 (713) 247-1837
San Antonio	William C. Sano	San Antonio Fire Department 115 Auditorium Circle San Antonio, TX 78205 (2 10) 299-7678

Washington

Seattle	Steven C. Bailey	Seattle Fire Department 301 Second Avenue South Seattle, WA 98104 (206) 386-1403

ARSON INFORMATION MANAGEMENT SYSTEMS (AIMS)

AIMS version 6 is an interactive, custom programmed, structured implementation of relational data files that contain arson-related information. It is "interactive" in that it interfaces with the various files searching for potential matches, provides on-screen help and verification of coded fields, and keeps the user informed of linkages between the various files and records. It also provides an environment in which the user can easily enter, retrieve, query, and maintain all information about a given case. The system's facilities help provide a high level of data quality at entry and edit time so that later searches will be as fruitful as possible.

Other capabilities provide potentials to agencies that wish to use them. For example, because of the system's linking abilities, complete property histories can be developed on target structures: sales histories, the chain of previous owners, policies, holding companies, and corporate officers all linked in a chain that can show patterns, and involvement in other cases and other properties with fire histories.

Attention has also been given to some of the important day-to-day case-management needs; an evidence tracking system and investigator assignment manager have been added.

For further information on AIMS contact:

Tom Minnich - Fire Prevention/Arson Control
FEMA/U. S . Fire Administration
National Emergency Training Center
16825 S. Seton Avenue
Emmitsburg, Maryland 21727

ARSON PUBLIC EDUCATION

Communities are combatting arson through the use of local public education campaigns. These campaigns alert the public to the seriousness of this crime and the need for community involvement.

Successful methods of educating the public include:

- News conferences held by recognized city leaders
- Press releases
- Newspaper articles and continuous coverage
- Arson specials for television
- Lectures to civic, school, and community groups.

The most common method to involve the public is the tipster/hotline program which includes a reward fund. The program and its identifiable slogan and symbol are advertised using the following:

- Television and radio public service announcements
- Billboards
- Signs on public transportation vehicles
- Newspapers and magazines
- Reward signs posted at sites of suspicious fires
- Flyers.

DIRECTORY OF EDUCATIONAL RESOURCES

Title	Type	Source
Adolescent Firesetter Handbook, Ages 14-18	Technical manual	U.S. Fire Administration Federal Emergency Management Agency P.O. Box 70274 Washington, DC 20024
Anatomy of an Arson	Videotape (VHS)	National Insurance Crime Bureau 10330 S. Roberts Road Pales Hills, IL 60465 (708) 430-2430 (708) 430-2446 (Fax)
Arson Burns Everybody	Slides	Rochester Fire Department Fire Investigation Attn: Captain William F. Kelly 365 Public Safety Building 150 S. Plymouth Avenue Rochester, NY 14614 (716) 428-7103
Arson Control Directory	Report	U .S . Fire Administration Federal Emergency Management Agency P.O. Box 70274 Washington, DC 20024
Arson Investigation	16mm film	National Fire Protection Association Fire Analysis and Research One Batterymarch Park Quincy, MA 02269 (6 17) 770-3000
Arson, It Should Burn You up	16mm film	Michigan Arson Prevention Committee Attn: John Wiechert Executive Director 200 Buhl Building Detroit, MI 48226 (313) 964-1435
Arson: Legal Considerations	Videotape with transcript	Emergency Education Network (EENET) National Emergency Training Center Emmitsburg, MD 21727

Title	Type	Source
Arson materials	Seminars, brochures, posters, signs, and door hangers	Colorado Advisory Committee on Arson Prevention c/o Western Insurance Information Service Attn: Shannon Kelly 6565 S. Dayton Street, Suite 2400 Englewood, CO 80111 (303) 790-0216
Arson-Our Most Costly **Crime**	16mm and videotape	Film Communicators Coronet/MT1 Film & Video 108 Wilmot Road Deerfield, IL 60015 (800) 777-2400 (708) 940-3640 (Fax)
Arson Prosecution: Issues and Strategies	Technical guide	U.S. Fire Administration Federal Emergency Management Agency P.O. Box 70274 Washington, DC 20024
Arson Resource Directory	Directory	U.S. Fire Administration Federal Emergency Management Agency P.O. Box 70274 Washington, DC 20024
Arson: W5 Series	All video formats (CC)- 12 minutes	CTV Television Network, Ltd. Attn: Marie Baccari, Manager Program Sales Department 42 Charles Street, E. Toronto, Ontario M4Y 1T5 Canada (4 16) 928-6095
An Arsonist Talks	16mm and videotape	Film Communicators Coronet/MT1 Film & Video 108 Wilmot Road Deerfield, IL 60015 (800) 777-2400 (708) 940-3640 (Fax)

Title	Type	Source
Bombs I *Bombs II* *Bombs III*	16mm and videotape 16mm and videotape 16mm and videotape	Film Communicators Coronet/MT1 Film & Video 108 Wilmot Road Deerfield, IL 60015 (800) 777-2400 (708) 940-3640 (Fax)
Bomb Search Procedures	Videotape	Film Communicators Coronet/MT1 Film & Video 108 Wilmot Road Deerfield, IL 60015 (800) 777-2400 (708) 940-3640 (Fax)
Crime of Arson	16mm and videotape	Film Communicators Coronet/MT1 Film & Video 108 Wilmot Road Deerfield, IL 60015 (800) 777-2400 (708) 940-3640 (Fax)
Curious Kids Set Fires	Ready-to-use materials (available in Spanish)	U.S. Fire Administration Federal Emergency Management Agency P.O. Box 70274 Washington, DC 20024
Essentials of Fire Fighting (3rd Edition)	Manual (French and Spanish available for 2nd Edition)	Fire Protection Publications Oklahoma State University Stillwater, OK 740780118 (800) 654-4055 (405) 744-5723

Establishing an Arson
   Task Force
	Technical guide	U.S. Fire Administration
		Federal Emergency Management
		   Agency
		P.O. Box 70274
		Washington, DC 20024

Title	Type	Source
Fire and Arson Investigators Field Index Directory	Directory	U.S. Fire Administration Federal Emergency Management Agency P.O. Box 70274 Washington, DC 20024
Firebugs	16mm and videotape	Film Communicators Coronet/MT1 Film & Video 108 Wilmot Road Deerfield, IL 60015 (800) 777-2400 (708) 940-3640 (Fax)
Firebugs and Your Job	Slide/tape	Factory Mutual Engineering and Research Corporation 1151 Boston Providence Turnpike P.O. Box 9102 Norwood, MA 02062 Attn: Publications Order Processing (617) 762-4300, Ext. 4681
Fire Cause Determination (1st Edition)	Manual	Fire Protection Publications Oklahoma State University Stillwater, OK 74078-0118 (800) 654-4055 (405) 744-5723
Firefighter on the Witness Stand	16mm and videotape	Film Communicators Coronet/MT1 Film & Video 108 Wilmot Road Deerfield, IL 60015 (800) 777-2400 (708) 940-3640 (Fax)
Fire in the Jail	16mm and videotape	Film Communicators Coronet/MT1 Film & Video 108 Wilmot Road Deerfield, IL 60015 (800) 777-2400 (708) 940-3640 (Fax)

Title	Type	Source
Fire Investigator on the Witness Stand	16mm and videotape	Film Communicators Coronet/MT1 Film & Video 108 Wilmot Road Deerfield, IL 60015 (800) 777-2400 (708) 940-3640 (Fax)
Get Low and Get Out	Film-7 minutes	Aetna Life and Casualty Company Corporate Communications RWAC Attn: Dolores Harper, Administrator 151 Farmington Avenue Hartford, CT 06156 (203) 273-2843
High Temperature Accelerant-Arson Fires Special Report	Report	U.S. Fire Administration Federal Emergency Management Agency P.O. Box 70274 Washington, DC 20024
Innovative Approaches for Arson Detection, Investigation and Prevention	Videotape with transcript	Emergency Education Network (EENET) National Emergency Training Center Emmitsburg, MD 21727
Interviewing & Counseling Juvenile Firesetters	16mm and videotape	Film Communicators Coronet/MT1 Film & Video 108 Wilmot Road Deerfield, IL 60015 (800) 777-2400 (708) 940-3640 (Fax)
Interviewing Witnesses	Videotape	Film Communicators Coronet/MT1 Film & Video 108 Wilmot Road Deerfield, IL 60015 (800) 777-2400 (708) 940-3640 (Fax)

Title	Type	Source
Introduction to Fire Cause Determination	Videotape	Film Communicators Coronet/MT1 Film & Video 108 Wilmot Road Deerfield, IL 60015 (800) 777-2400 (708) 940-3640 (Fax)
Kids Playing with Fire: 7he Clarke Family Tragedy	16mm and videotape	Film Communicators Coronet/MT1 Film & Video 108 Wilmot Road Deerfield, IL 60015 (800) 777-2400 (708) 940-3640 (Fax)
Killer Arson	16mm and videotape	Film Communicators Coronet/MT1 Film & Video 108 Wilmot Road Deerfield, IL 60015 (800) 777-2400 (708) 940-3640 (Fax)
The National Fire Drill	All video formats- 30 minutes (available in French)	CTV Television Network, Ltd. Attn: Marie Baccari, Manager Program Sales Department 42 Charles Street, E. Toronto, Ontario M4Y lT5 Canada (416) 928-6095
Our Town Is Burning Down	Film- 18 minutes	Aetna Life and Casualty Company Corporate Communications RWAC Attn: Dolores Harper, Administrator 151 Farmington Avenue Hartford, CT 06156 (203) 273-2843
Planning for Bomb Threats	Videotape	Film Communicators Coronet/MT1 Film & Video 108 Wilmot Road Deerfield, IL 60015 (800) 777-2400 (708) 940-3640 (Fax)

Title	Type	Source
A Pocket Guide to Arson Investigation	Handbook	Factory Mutual Engineering and Research Corporation 1151 Boston Providence Turnpike P.O. Box 9102 Norwood, MA 02062 Attn: Publications Order Processing (617) 762-4300, Ext. 4681
Preadolescent Firesetter Handbook, Ages O-7	Technical manual	U.S. Fire Administration Federal Emergency Management Agency P.O. Box 70274 Washington, DC 20024
Preadolescent Firesetter Handbook, Ages 7-13	Technical manual	U.S. Fire Administration Federal Emergency Management Agency P.O. Box 70274 Washington, DC 20024
Protection Against Incendiary Fires	Pamphlet	Factory Mutual Engineering and Research Corporation 1151 Boston Providence Turnpike P.O. Box 9102 Norwood, MA 02062 Attn: Publications Order Processing (617) 762-4300, Ext. 4681
Report of the Operation Urban Wildfire Task Force	Report	U.S. Fire Administration Federal Emergency Management Agency P.O. Box 70274 Washington, DC 20024
Rochester Fire Investigation Overview	Videotape (VHS)	Rochester Fire Department Fire Investigation Attn: Captain William F. Kelly 365 Public Safety Building 150 S. Plymouth Avenue Rochester, NY 14614 (716) 428-7103

Title	Type	Source
Rochester Fire-related Youth Overview	Videotape (VHS)	Rochester Fire Department Fire Investigation Attn: Captain William F. Kelly 365 Public Safety Building 150 S. Plymouth Avenue Rochester, NY 14614 (716) 428-7103
Rural Arson Control	Report	U.S . Fire Administration Federal Emergency Management Agency P.O. Box 70274 Washington, DC 20024
Safe at Home	Brochure	Allstate Insurance Company Allstate Plaza Northbrook, IL 60062 (708) 390-9650
Smoke and Fire-TWO Steps to Survival	Film- 19 minutes	Aetna Life and Casualty Company Corporate Communications RWAC Attn: Dolores Harper, Administrator 151 Farmington Avenue Hartford, CT 06156 *(203) 273-2843*
Torch of Destruction	All video formats- 11 minutes	CTV Television Network, Ltd. Attn: Marie Baccari, Manager Program Sales Department 42 Charles Street, E. Toronto, Ontario M4Y lT5 Canada (416) 928-6095
U.S. Arson Trends and Patterns-1991	Report (updated annually) 11 units of 35mm slides, dealing with detection, evidence, investigation, prosecution, and trial procedures and containing audio cassettes and manuals	National Fire Protection Association Fire Analysis and Research One Batterymarch Park Quincy, MA 02269 (617) 770-3000

DIRECTORY OF EDUCATIONAL RESOURCES (Continued)

Title	Type	Source
A View of Management in Fire Investigation Units, Volumes I and II	Report	U.S. Fire Administration Federal Emergency Management Agency P.O. Box 70274 Washington, DC 20024
Why You Should See Red over Arson	Brochure	Allstate Insurance Company Allstate Plaza Northbrook, IL 60062 (708) 390-9650
Winning the War on Arson	Film- 15 minutes	Aetna Life and Casualty Company Corporate Communications RWAC Attn: Dolores Harper, Administrator 151 Farmington Avenue Hartford, CT 06156 (203) 273-2843

PUBLIC ARSON EDUCATION RESOURCES

Insurance Industry

The insurance industry has developed effective tools for educating the public. The resources listed below can provide communities, organizations, and public safety agencies with literature, planning manuals, audiovisual materials, media campaign materials, and-at times-seed money for programs.

Aetna Life and Casualty
Arson and Fraud Unit
151 Farmington Avenue
Hartford, CT 06156
Contact: John L. Swedo, Asst. Vice President
(800) 323-8648

Alliance of American Insurers
1501 Woodfield Road, Suite 400 West
Schaumburg, IL 60173-4980
Contact: Dean M. Moffitt, Vice President,
   Personal Lines Division
(708) 330-8526
(708) 330-8602 (Fax)

Allstate Insurance Company
Allstate Plaza F3
Northbrook, IL 60062
Contact: David B. Warstler, Senior Corporate
   Relations Manager
(708) 402-2908

Factory Mutual Engineering and Research
   Corporation
1151 Boston Providence Turnpike
P.O. Box 9102
Norwood, MA 02062
Contact: Donald C. Garner, Arson Coordinator
(617) 762-4300

ITT Hartford Insurance Group
Hartford Plaza
Hartford, CT 06115
Contact: Melissa H. Engel
(203) 547-4711

Insurance Committee for Arson Control
110 William Street
New York, NY 10038
Contact: Richard Gilman, Executive Director
(2 12) 669-9245
(212) 732-1916 (Fax)

Insurance Information Institute
110 William Street
New York, NY 10038
Contact: Gordon C. Stewart
(2 12) 669-9200

New York Property Insurance Underwriters
   Association, Inc.
100 William Street, 4th Floor
New York, NY 10038
Contact: Kenneth Lang, Asst. Vice President,
   Claims
(212) 208-9813

Professional Insurance Agents
400 N. Washington Street
Alexandria, VA 22314
Contact: James Quiggle, Director, Public
   Relations
(703) 836-9340

SAFECO Insurance Companies
Safeco Plaza
Seattle, WA 98185
Contact: Gordon C. Hamilton, Vice President
   of Public Relations
(206) 545-5705

St. Paul Fire and Marine Insurance Company
385 Washington Street
St. Paul, MN 55102
Contact: Karen Himley, Vice President,
   Communications
(612) 221-7911

State Farm Fire and Casualty Company
112 E. Washington Street
Bloomington, IL 61701
Contact: David Stuart, Claim Consultant
(309) 766-2983

Western Insurance Information Service
Colorado Advisory Committee on Arson
   Prevention
6565 S. Dayton Street, Suite 2400
Englewood, CO 80111
Contact: Shannon Kelly
(303) 790-0216

PART I

B. Programs and Resources for Investigation/Prosecution of Arson

PROGRAMS AND RESOURCES
FOR INVESTIGATION/PROSECUTION OF ARSON

The causes of many arsons go undetected because fires are often believed to be accidental. Many arsonists are never caught and even if they are arrested, arson is difficult to prove in court. Evidence is often destroyed in the fire; thus, most cases are based on circumstantial or indirect evidence.

The low success rate of arson prosecutions can be attributed to the following reasons:

- Firefighters first on scene were untrained to detect the cause of the fire

- Fire investigators were either too few in number, or inadequately trained

- Prosecutors, inexperienced with arson cases or overburdened with caseloads, avoided the difficult arson cases.

Communities and organizations, alarmed by the arson problem, have sought to improve the training of firefighters, investigators, and prosecutors. The following list identifies some of these education and training programs.

Federal programs are listed first. These are followed by State programs and programs for prosecuting attorneys.

FEDERAL TRAINING PROGRAMS

Federal Training Efforts

National Fire Academy

The National Fire Academy, Emmitsburg, Maryland, offers the following training courses in arson detection and investigation:

"Fire/Arson Investigation"

The course addresses skills in conducting fire investigations. Arson bum buildings are used to demonstrate techniques in legal fire investigation to assist in successful arson prosecutions. Students completing the course will be equipped to identify fire origin and cause; conduct legal, technically correct investigations; and properly pursue the case through the judicial system.

The course includes both classroom learning and "hands-on" training, acquired from bum building scene examinations and case development.

This course is recommended for credit by the American Council on Education at the upper-division baccalaureate or graduate level for 3 credit hours in Arson Investigation, Fire Science, Criminal Justice, or Insurance.

Course Duration: 10 days

Further information on National Fire Academy courses may be obtained from:

Superintendent
National Fire Academy
16825 S. Seton Avenue
Emmitsburg, MD 21727
(301) 447-1117

Program Chair
National Fire Academy
16825 S. Seton Avenue
Emmitsburg, MD 21727
(301) 447-1086

71

FEDERAL TRAINING PROGRAMS (Continued)

Bureau of Alcohol, Tobacco, and Firearms

The Bureau of Alcohol, Tobacco, and Firearms (ATF), conducts training sessions focusing on arson-for-profit. The curriculum covers: arson-for-profit investigation techniques, laws related to arson, and the role of the crime lab in arson investigation. Information on these training sessions may be obtained from:

Chief, Training Division
Bureau of Alcohol, Tobacco, and Firearms
650 Massachusetts Avenue, N. W.
Washington, DC 20226
(202) 927-7920

STATE TRAINING

ALABAMA
Mr. W.L. Langston, Executive Director
Alabama State Fire College
2015 McFarland Avenue East
Tuscaloosa, AL 35405
(205) 759-1508

Fire/Arson Detection
Basic Fire Arson Investigation
Advance Fire Arson Investigation

ALASKA
Mr. Mark Barker
Administrator
Fire Service Training
5700 E. Tudor Road
Anchorage, AK 99507
(907) 269-5789

Fire Cause and Origin Investigation

ARIZONA
Mr. Bob Costello
Director of Fire Training
Department of Building & Fire Safety
Office of the State Fire Marshal
1540 W. Van Buren
Phoenix, AZ 85007
(602) 255-4964

Fire/Arson Detection

ARKANSAS
Mr. Sheldon Richardson, Director
Arkansas Fire Academy
SAU-Tech
P.O. Box 3499
East Camden, AR 71701
(501) 574-1521

Fire/Arson Detection
Fire/Arson Investigation

CALIFORNIA
Mr. Art Cota, Supervisor
State Fire Training
California Fire Service Training
 and Education System
7171 Bowling Drive, Suite 600
Sacramento, CA 95823
(916) 427-4204

Fire and Arson Detection
Fire Cause and Origin Determination
Criminal and Legal Procedures
Field Case Studies
Techniques of Fire Investigation

COLORADO

Mr. Dean W. Smith, Director   Fire/Arson Detection
Colorado Division of Fire Safety
700 Kipling, Suite 1200
Denver, CO 80215
(303) 239-4463

CONNECTICUT

Mr. Jeffrey J. Morrissette, Acting   Fire/Arson Detection
 Administrator
294 Colony Street, Bldg. #5
Meriden, CT 06450
(203) 238-6587

DELAWARE

Mr. Louis J. Amabili, Director   Fire/Arson Detection
Delaware State Fire School
RD 2, Box 166
Dover, DE 19901
(302) 739-4773

FLORIDA

Mr. Fred C. Stark, Bureau Chief   Arson Investigation
Florida State Fire College   Cause and Origin
11655 Northwest Gainesville Road   Chemistry for Arson Investigators
Ocala, FL 34482-1486   Latent Investigation
(904) 732-1330   Legal Issues

GEORGIA

David Pritchett   Arson Detection
Acting Superintendent   Special Courses on Fire Investigation
Georgia Fire Academy
1000 Indian Springs Drive
Forsyth, GA 31029
(912) 993-4670

HAWAII

August Range   Various Arson-related Courses
Coordinator, Hawaii State Fire Council
3375 Koapaka Street, Suite H425
Honolulu, HI 96819
(808) 381-7748

74

IDAHO

Mr. Clare D. Harkins, Director
Fire Service Training
State Division for Vocational Education
650 W. State Street, Room 324
Boise, ID 83720
(208) 334-3211

Fire/Arson Detection
Special Courses on Fire Investigation

ILLINOIS

Mr. Gerald Monigold, Director
Fire Service Institute
University of Illinois
F.S.I. Building
11 Gerty Drive
Champaign, IL 61820

Fire/Arson Investigation-Level I,
  II, III
Arson Detection
Arson Awareness
Juvenile Fire Setter

INDIANA

Mr. Ivan Nevil, Director of
  Education and Training
402 W. Washington Street
Room 241E
Indianapolis, IN 46204
(3 17) 232-2222

Fire/Arson Investigation
Fire/Arson Detection

IOWA

Mr. George Oster, Supervisor
Fire Service Institute
Iowa State University
Haber Road
Ames, IA 50011-3100
(515) 294-6817

Fire and Arson Detection
  (Through State Fire Marshal)

KANSAS

Mr. Alan Walker, Director
University of Kansas
Fire Service Training
645 New Hampshire Avenue
Lawrence, KS 66044
(913) 864-4467 or 4873

Basic Fire Arson Detection
Advanced Fire Arson Investigation
Cause and Determination
NFPA Fire Investigator Certification
Explosive Investigations

KENTUCKY

Mr. Bill Denton, Supervisor
Public Service/Short-Term Training Unit
Office of Vocational Education,
    Kentucky Department of Education
Capitol Plaza Tower, 20th Floor
Frankfort, KY 40601
(502) 564-2326

Arson Detection
Arson Investigation

LOUISIANA

Mr. Thomas Hebert, Director
LSU Fireman Training Program
6868 Nicholson Drive
Louisiana State University
Baton Rouge, LA 70820
(504) 766-0600

Cause and Origin Investigation
Arson Detection
Fire/Arson Investigation

MAINE

Mr. Stephen G. Hasson, State
    Administrator
Fire Training & Education
Southern Maine Technical College
Fort Road
South Portland, ME 04106
(207) 799-7303

Fire Investigation-Level I and II

MARYLAND

Mr. Steven T. Edwards, Director
Maryland Fire & Rescue Institute
Fire Service Building
University of Maryland
College Park, MD 20742
(30 1) 454-2416

Fire/Arson Detection
Fire Investigation

MASSACHUSE'ITS

Mr. Stephen Coan, Director
Massachusetts Firefighting Academy
State Road, P.O. Box 1025
Stowe, MA 01775
(508) 562-1400

Fire Investigation

MICHIGAN
Michigan Firefighter's Training Council
7150 Harris Drive
Lansing, MI 48913
(517) 322-1922

Fire/Arson Detection
Fire Investigation Course
Basic Fire Arson Course
Advanced Fire Arson Course

MINNESOTA
Mr. Adam D. Piskura
Fire Service Training Director
Minnesota Technical College System
F.I.R.E. Center
550 Cedar Street
St. Paul, MN 55101
(612) 296-6516

Fire and Arson Investigation-Level I,
II, III

MISSISSIPPI
Mr. William Warren, Director
Mississippi Fire Academy
Route 10, Box 295
Jackson, MS 39208
(601) 932-2444

Fire Investigator

MISSOURI
Mr. Bruce R. Piringer, Director
Fire & Rescue Training Institute
205 Lewis Hall
University of Missouri-Columbia
Columbia, MO 65211
(3 14) 882-4735

Fire Cause Determination

MONTANA
Mr. Seldon "Butch" Weedon, Director
Montana Fire Service Training School
2100 16th Avenue, S.
Great Falls, MT 59405
(406) 761-7885

Fire/Arson Detection
Advanced Fire Investigation Seminar

NEBRASKA
Program Manager
Nebraska Fire Service
3721 W. Cuming
Lincoln, NE 68524-1896
(402) 471-2803

Fire/Arson Detection

NEVADA
Mr. Robert Hitchens
State Training Director
Fire Service Training Section
State Fire Marshal Division
Capitol Complex
Carson City, NV 89710
(702) 687-4290

Fire/Arson Investigation
Fire/Arson Detection

NEW HAMPSHIRE
Mr. Barry Bush, Chief
New Hampshire Fire Standards and
   Training Commission
91 Airport Road
Concord, NH 03301
(603) 271-2661

Arson Investigation
N.H. Citation Authority Course

NEW JERSEY
Training Academy
New Jersey Division of Criminal
   Justice
P.O. Box 178
Fort Dix, NJ 08640-0178

Basic Course for Arson Investigators

NEW MEXICO
Mr. Bob Baca, State Fire Marshal
New Mexico State Fire Marshal's Office
P.O. Drawer 1269
Santa Fe, NM 87504-1269
(505) 827-3721
(800) 244-6702

Cause and Origin Investigation
Fire/Arson Investigation

NEW YORK
Mr. Richard Harris, Director of
   Programming & Education
Office of Fire Prevention and Control
New York State Department of State
162 Washington Avenue
Albany, NY 12231
(5 18) 474-6746

Cause and Origin Determination
Fire/Arson Investigation
Specialized Courses on Fire Investigation

NORTH CAROLINA

Mr. Ken Farmer, Director
Fire Training Services
Department of Community Colleges
Education Building, Room 176
Raleigh, NC 27611
(919) 733-3345

Fire/Arson Detection
Firefighters' Role in Arson Detection

OHIO

Mr. A. Gregory Drew, Superintendent
Ohio Fire Academy
Ohio Division of State Fire Marshal
8895 E. Main Street
Reynoldsburg, OH 43068
(614) 864-5510

Fire/Arson Detection

OKLAHOMA

Ms. Nancy Trench, Assistant Director
Fire Service Training
Fire Building
Oklahoma State University
Stillwater, OK 74078-0114
(405) 744-5727

Fire Cause Determination and Investigation
Fire Arson Detection I

OREGON

Mr. Dan Bolis, Director
State Fire Training
4760 Portland Road, N.E.
Salem, OR 97305-1760
(503) 378-3473

Fire Cause Investigation

PENNSYLVANIA

Mr. Timothy L. Dunkle, Administrator
Pennsylvania State Fire Academy
1150 Riverside Drive
Lewistown, PA 17044
(717) 248-1115

Fire/Arson Detection
Arson Detection and Fire Investigation

RHODE ISLAND
Mr. Henry Serbst
Chief of Investigations
Rhode Island State Fire Marshal's Office
272 W. Exchange Street
Providence, RI 02903
(40 1) 277-2335

Basic Fire Investigation
Electrical Fire Investigation
Fatal Fire Investigation
Mock Trials

SOUTH CAROLINA
Mr. Preston Cantrell, Acting Director
South Carolina Fire Academy
2920 Fire Academy Road
West Columbia, SC 29170
(803) 822-5380

Fire/Arson Detection

SOUTH DAKOTA
Mr. Allen L. Christie, Director
Fire Service Training
118 W. Capitol
Pierre, SD 57501
(605) 773-3876

Fire/Arson Detection

TENNESSEE
Mr. Wallace Burke, Director
Tennessee State Fire School
1303 Old Fort Parkway
Murfreesboro, TN 37129
(615) 898-8010

Fire/Arson Detection

TEXAS
Mr. Charles Page, Division Head
Fire Protection Training
Texas Engineering Extension Service
Texas A & M University
College Station, TX 77843-8000
(409) 845-7641

Fire/Arson Detection
Arson Investigation
Fire Cause Investigation

UTAH
Mr. Steve Lutz, Director
State Fire Service Training
Utah Valley Community College
800 West 1200 South
Orem, UT 84058
(801) 222-8000, Ext. 508

Fire Cause Determination

VERMONT
Mr. Wayne Babcock, Supervisor
Fire Service Training Program
VT State Firefighter Association
P.O. Box 85
Chelsea, VT 05038
(802) 685-3112

Fire/Arson Investigation

VIRGINIA
Mr. Kenneth Sharp, Executive Director
Virginia Department of Fire Programs
2807 Parham Road, Suite 200
Richmond, VA 23294
(804) 527-4236

Fire/Arson Detection

WASHINGTON
Mr. John Anderson, Director
Fire Service Training
Commission for Vocational Education
Airdustrial Park
Building 17, Mail Stop LS-10
Olympia, WA 98504
(206) 753-5679

Fire/Arson Detection
Arson Investigation
Advance Arson Investigation
Arson Investigation/Police Science
Advance Arson Investigation/Police Science

WEST VIRGINIA
Mr. Everett Perkins, Program Leader
Fire Service Extension
P.O. Box 6610
Monongahela Boulevard
West Virginia University
Morgantown, WV 265066610
(304) 293-2106

Fire Cause Determination

WISCONSIN
Mr. David J. Brooks
Fire Education and Training
310 Price Place
P.O. Box 7874
Madison, WI 53707
(608) 266-7994

Fire/Arson Detection

WYOMING
Ms. Nancy Eagle
Training Coordinator
Fire Prevention and Electrical Safety
2301 Central Avenue
Barret Building, 4th Floor
Cheyenne, WY 82002
(307) 777-7907

Fire and Arson Detection

PROSECUTOR TRAINING

Legal prosecution plays an important role in deterring and punishing arsonists. Prosecutors, unfamiliar with arson, have traditionally been reluctant to pursue arsonists. However, prosecutors are now better informed about arson, and have become more aggressive in taking arsonists to court. Also, many prosecutors serve on arson task forces and agencies which devise anti-arson strategies. State district attorney organizations are spearheading training programs which will help prosecutors be more effective in prosecuting arsonists.

DIRECTORY OF PROSECUTORS FOR ARSON TRAINING PROGRAMS

National College of District Attorneys
University of Houston Law Center
Houston, TX 77204-6380
Contact: Roger Kane
(7 13)747-6232

California District Attorneys' Association
1414 K Street, Suite 300
Sacramento, CA 95814
Contact: Mike Yadon
(916)443-2017

PART II

Other Anti-Arson Resource Organizations

INTERNATIONAL ASSOCIATION OF ARSON INVESTIGATORS (IAAI)

OFFICERS

President
Jean-Claude Cloutier
ICPB
365 Evans Avenue, 4th Floor
P.O. Box 919, Station "U"
Toronto, Ontario
M8Z 5P9 Canada

1st Vice President
Robert Whitemore
Chief Investigator
Robins, Kaplan, Miller, et al.
2800 LaSalle Plaza
800 LaSalle Avenue
Minneapolis, MN 55402-2015

2nd Vice President
John Barracato
38 Orchard Hill Drive
South Windsor, CT 06074

DIRECTORS

Michael P. Carrocci
Fire Investigator
SEA Investigations Div., Inc.
7349 Worthington-Galena
Columbus, OH 43085

Ray J. Castaldi
Marple County Township Pol.
8 Grove Lane
Broomall, PA 19008

Alan L. Clark
Director, Invest. Serv.
Grinnell Mut. Reinsurance
Int. 80 and Route 146
Grinnell, IA 50112

Kevin T. Cunningham
General Manager
Cunningham Investigative Services, Inc.
P.O. Box 1279
Snellville, GA 30278

William S. Daniel
Attorney, Daniel Law Office
5610 W. Main Street
Belleville, IL 62223

Robert E. Downer
State Farm Ins. Co.
2500 Memorial Boulevard
Murfreesboro, TN 37129

V. Ray Eastman
Assistant Director
North Carolina State Bureau of Invest.
P.O. Box 29500
Raleigh, NC 27626

Philip Horbert
Special Agent in Charge
Bureau of Alcohol, Tobacco and Firearms
Explosives Tech. Branch
650 Massachusetts Avenue, N. W.
Washington, DC 20226

Larry E. Julian
Sergeant
Michigan State Police
Flint Post
G-3478 Corunna Road
Flint, MI 48504

David E. Marsh
Div. Clm. Supt.
State Farm Ins. Co.
P.O. Box 1045
Beckley, WV 25802-1045

Judith Maydew
Branch Manager
Underwriters Adjustment
803 T D Tower
10205-101 Street, Edmonton Centre
Edmonton, Alberta
T5J 221 Canada

Joseph D. Molnar
Sr. Loss Analyst
Interscience, Inc.
5440 Beaumont Center Boulevard, #490
Tampa, FL 34625

Gerard Naylis
Arkwright
Account Engineer
201 Route 17, N.
Rutherford, NJ 07070

William C. Vielhauer
Investigator
North Country Ins. Co.
P.O. Box 237
921 State Street
Ogdensburg, NY 13669

Jack Yates
President/Owner
Yates and Associates, Inc.
2017 S. Elm Place, Suite 108
Broken Arrow, OK 74012

CHAPLAIN

Rev. Fr. David W. Arnold
17 Alder Court
Kingston, NY 12401-6901

COUNSEL

Terry-Dawn Hewitt
Brownlee & Fryett
Barristers & Solicitors
2300-10104 - 103 Avenue
Edmonton, Alberta T5J 3X7 Canada

EXECUTIVE OFFICE

Benny King
Executive Secretary
300 S. Broadway, Suite 100
St. Louis, MO 63102
(314) 621-1966
(314) 621-5125 (Fax)

International Association of Arson Investigators
CHAPTER PRESIDENTS AND SECRETARIES

State	Contact	Address
Alabama	Rodney Brown President	ALFA Mutual Insurance Company P.O. Box 66 Heflin, AL 36264 (205) 463-2295
	Gerald Bartig Secretary	11207 Woodcrest Drive, S.E. Huntsville, AL 35803 (205) 883-0429
Alaska	Michael Donovan President	12408 Winterpark Circle Eagle River, AK 99577 (907) 852-6111
	Claude Adams Secretary	Anchorage Fire Department 11244 Fireball Street Eagle River, AK 99577 (907) 694-2675
Alberta	Gwynn Edwards President and Secretary	#505 10333 South Port Road, S.W. Calgary, Alberta T2W 3X6 Canada (403) 258-3677
Arizona	Eric Cooper President	Pima County Sheriff P.O. Box 910 Tucson, AZ 85702 (602) 746-3791
	Chip Shilosky Secretary	3820 Cholla Drive Lake Havasu, AZ 86403 (602) 453-3313
Arkansas	Greg Rudell President	901 Kenwood Benton, AR 72015 (501) 371-4769
	Milton Dillingham Secretary	Stuttgart Fire Department 512 S. Main Street Stuttgart, AR 7260 (501) 673-3539

State	Contact	Address
British Columbia	Daryl Driemel President	Saanich Fire Department 780 Vernon Avenue Victoria, British Columbia VSX 2W7 Canada (604) 388-5531
	Dan Lemieux Secretary	4595 Meadowbank Close North Vancouver, British Columbia V7K 2Ll Canada (604) 988-7421
California	Mark E. Johnson President	1122 Lincoln Avenue Orange, CA 92665 (714) 283-0986
	Tom Kuczynski Secretary	Fresno Fire Department 450 M Street Fresno, CA 93721 (209) 498-4253
Colorado	Donald F. Peak President	P.O. Box 27297 Denver, CO 80227 (303) 936-8435
	Jim Kalahar Secretary	10544 W. Marlowe Place Littleton, CO 80127 (303) 972-6721
Conneticut	Pam Craig-White President	12 Sylvan Valley Road Meriden, CT 06450 (203) 273-2108
	Dennis Flynn Secretary	9 Woodridge Road West Haven, CT 06516 (203) 932-4262
Delaware	Edward Hojnicki, Jr. President	1331 Maple Street Wilmington, DE 19805 (302) 571-4581

State	Contact	Address
Delaware (Cont.)	Robert Montgomery, Jr. Secretary	159 Cook Court Smyrna, DE 19977 (302) 739-4447
DC/Maryland	Dan Boeh President	Bureau of Alcohol, Tobacco and Firearms 31 Hopkins Plaza, Room 1120 Baltimore, MD 21201 (301) 962-4115
	Sue Beck Secretary	Bureau of Alcohol, Tobacco and Firearms 1401 Research Boulevard Rockville, MD 20850 (301) 294-0420
Florida	Larry Weintraub President	275 N.W. 2nd Street Miami, FL 33128 (305) 350-7832
	Robert R. Gentile	2302 N. Wallen Drive Lake Park, FL 33410 (407) 746-1192
Georgia	Mike Weaver President	433 Pecan Circle Forsyth, GA 31029 (912) 474-8411
	Frank H. Carter Secretary	Inserv South, Inc. 28 Perimeter Center, E., Suite 100 Atlanta, GA 30346 (404) 292-7875
Idaho	Terry Edwards President	P.O. Box 50220 Idaho Falls, ID 83405 (208) 529-1495
	Ben Estes Secretary	Pocatello Fire Department 408 E. Whitman Pocatello, ID 83201 (208) 234-6203

State	Contact	Address
Jllinois	Reuben B. King President	Illinois Fire Services Institute #ll Gerty Avenue Champaign, IL 61820 (217) 333-4215
	Carl Dropka Secretary	2331 S. Des Plaines Avenue North Riverside, IL 60546 (708) 447-1981
Indiana	Karl J. Benz President	State Farm Fire & Casualty Co. P.O. Box 2308 Clarksville, IN 47131 (812) 288-1324
	Steve Cook Secretary	615 N. Dequincy Street Indianapolis, IN 46201 (317) 351-9804
I o w a	Robert D. Burn, Jr. President	344 3rd Avenue, S. Clinton, IA 52732 (319) 424-0125
	Roger Heglund Secretary	2302 N.W. 9th Street Ankeny, IA 50021 (515) 965-6469
Israel	Shalom Tsaroom President	29 Altman Street Pisgat Zeev Jerusalem, 97552 Israel 011-972-2-309435
	Amnon Ramon Secretary	2 Abarbanel Street Jerusalem, 92425 Israel 011-972-2-243068
K a n s a s	Michael Schlatman President	P.O. Box 25852 Overland Park, KS 66225 (913) 469-1113

State	Contact	Address
Kansas (Cont.)	Donald L. Kimsey Secretary	1404 Elizabeth Winfield, KS 67156 (316) 221-0530
Kentucky	William Lilly President	Lexington Fire Department 200 E. Main Street Lexington, KY 40507 (606) 254- 1120
	Mike Barry Secretary	3421 Boston Road Lexington, KY 40503 (606) 223-1644
Louisiana	James Duncan President	9088 Shadow Bluff Avenue Denham Springs, LA 70726 (504) 664-8598
	Aron K. Hoyt Secretary	P.O. Box 4462 Pineville, LA 71361-4462 (3 18) 449-5666
Massachusetts	Susan Mackay President	Hanover Insurance P.O. Box 268 Tewkesbury, MA 01876 (508) 851-7600
	D. Eileen Silvia Secretary	Rhode Island Fair Plan 44 Helm Street Jamestown, RI 02835 (800) 851-8978
Michigan	James A. Labuhn President	Mt. Clemens Fire Department 2 Dickinson Mt. Clemens, MI 48043 (3 13) 469-6840
	Gwen White-Erickson Secretary	Michigan State Police 42145 W. Seven Mile Road Northville, MI 48167 (313) 473-1114

State	Contact	Address
Minnesota	Mary Nachbar, Fire Marshal President	450 N. Syndicate, Suite 285 St. Paul, MN 55102 (612) 643-3086
	Dave Strege Secretary	State Farm Insurance 1500 West Highway 36 St. Paul, MN 55161 (612) 631-4455
Mississippi	W. "Ronnie" T. Baughn President	P.O. Box 675 Ocean Springs, MS 39702 (205) 690-2338
	Thomas O. Saffle, Jr. Secretary	809 Spring Lake Drive Terry, MS 39170 (601) 373-4364
Missouri	Joseph Reichwein President	P.O. Box 846 Lake Ozark, MO 65049 (3 14) 365-2462
	David I. Snarr Secretary	125 Madison Macon, MO 63552 (816) 385-6436
Montana	Creighton Sayles President	Eagle Investigations, Inc. 9397 Upper Miller Creek Missoula, MT 59803 (406) 251-4260
	Bob Knudson Secretary	Helena Fire Department 316 N. Park Helena, MT 59601 (406) 442-9920, Ext. 470
Nebraska	Charles Beachcamp President	7134 Starr Street Lincoln, NE 68505 (402) 471-7791

State	Contact	Address
Nebraska (Cont.)	Charles H. Hoffman Secretary	2115 Park Drive Grand Island, NE 68801 (308) 384-5296
Nevada.	Joe Skaarup President	2626 E. Carey Avenue North Las Vegas, NV 89030 (702) 649-4222
	Gary Tecklenburg Secretary	2626 E. Carey Avenue North Las Vegas, NV 89030 (702) 649-4222
	Sonja Drase Correspondence & Contact	1202 S. Martin Luther King Boulevard Las Vegas, NV 89102 (702) 388-8787
New Hampshire	Allen W. Britton President	State of New Hampshire 34 Beaver Street Nashua, NH 03063 (603) 271-3294
	George E. Sykes Secretary	Lebanon Fire Department 3 Avon Avenue Lebanon, NH 03766-1102 (603) 448-1212
New Jersey	John Hannah President	120 Grandview Drive Woodstown, NJ 08098 (609) 877-0800
	David Kircher Secretary	c/o Monmouth County Fire Academy 1027 Highway 33 Freehold, NJ 07728 (908) 621-5694
New Mexico	Ted J. Alford President	Bureau of Alcohol, Tobacco and Firearms 517 Gold S.W. Albuquerque, NM 87102 (505) 766-2271

State	Contact	Address
New Mexico (Cont.)	Chris Hill Secretary	Box 601 Pie Town, NM 87827 (505) 835-4091
New South Wales, Australia	Alan McFarlane President	GIO Australia, Level 2 146 Marsden Street Parramatta, NSW 2150 Australia 011-61-02-891-7601
	Ross Blowers Secretary	CIC Insurance 6 O'Connell Street Sydney, NSW 2000 South Australia 011-61-02-224-5610
New York	Gary Printy President	170 Devonshire Drive Rochester, NY 14625 (716) 473-3162
	Craig W. Corey Secretary	248 Pinebrook Drive Rochester, NY 14616 (716) 473-3162
New Zealand	Wayne R. Lawrence President	18 Sydney Street Palmerston North New Zealand 011-64-6-358-4242
	John Kelliher Secretary	c/o P.O. Box 2148 Auckland New Zealand 011-64-9-798-269
North Carolina	David H. Campbell President	1150 Maynard Road, Suite 101 Cary, NC 27511 (9 19) 469-5707
	Nancy H. Johnson, AIC Secretary	14039 Woody Point Road Charlotte, NC 28278 (704) 329-2520

State	Contact	Address
Nova Scotia	Janice Myra President	Zurich-Canada Metropolitan Place 99 Wyse Road, 12th Floor Dartmouth, Nova Scotia B3A 4S5 Canada (902) 866- 1096
	Leonard M. Stevens Secretary	Marsh Adjustment Bur. Ltd. 224 Dufferin Street Bridgewater, Nova Scotia B4V 2G7 Canada (902) 543-3222
Ohio	James E. Harting President	State Fire Marshal's Office 8895 E. Main Street Reynoldsburg, OH 43068 (614) 752-7103
	Michael K. Simmons Secretary	8895 E. Main Street Reynoldsburg, OH 43068 (614) 752-7104
Oklahoma	David Wiist President	925 E. Second Edmond, OK 73034 (405) 348-1585
	Mark Keim Secretary	2300 General Pershing Boulevard Oklahoma City, OK 73106 (405) 297-3321
Ontario	Richard Walters President	1336 Sandhill Drive Lancaster, Ontario L9G 4V5 Canada (4 16) 648-5522
	Barry S. (Tim) Burch Secretary	252 Dundas Street Cambridge, Ontario NlR 5T3 Canada (519) 623-1910

State	Contact	Address
Oregon	Charles Powers President	55 S.W. Ash Street Portland, OR 97213 (503) 823-3791
Pennsylvania	Dave Harms Secretary	55 S.W. Ash Street Portland, OR 97213 (503) 823-3700
	Raymond Cobb President	RD #2, Box 55F Hallstead; PA 18822 (7 17) 879-4609
Rhode Island	Michael Moonblatt Secretary	8230 Old York Road Elkins Park, PA 19117 (215) 887-1561
	Arthur Solvang President	3 Cheshire Drive Barrington, RI 02806 (401) 277-2335
	Capt. Charles Hall Secretary	123 Setian Lane West Warwick, RI 02893 (401) 461-4227
Saskatchewan	Wayne H. Miner President	Saskatchewan Gov. Ins. 2260 11th Avenue Regina, Saskatchewan S4P OJ9 Canada (306) 566-6049
	Doug Sanders Secretary	c/o Fire Comm. Branch 3130 - 8th Street, E. Saskatoon, Saskatchewan S7H OW2 Canada (306) 933-5063
South Carolina	Rusty Horton President	859 Woodberry Road Lexington, SC 29073 (803) 781-7652

State	Contact	Address
South Carolina (Cont.)	Wilmon L. Hutto, Jr. Secretary	3720 Mineral Springs Road Lexington, SC 29073 (803) 356-4345
South Dakota	Jim Gordon President	P.O. Box 975 Pierre, SD 57501 (605) 224-4981
	Helen King Secretary	RR1, Box2B Blunt, SD 57522 (605) 773-4311
Tennessee	Robert Downer President	State Farm Insurance Company 2500 Memorial Drive Murfreesboro, TN 37129 (615) 898-6000
	Gary 0. Chappell Secretary	545 Hillside Lane Gallatin, TN 37066 (615) 741-3030
Texas	Gerald R. Brown President	P.O. Box 3133 Port Arthur, TX 77643 (409) 962-1513
	John R. Rauch Secretary	2713 Forest Trail Temple, TX 76502-2545 (409) 774-9401
Utah	Jeffrey Long President	315 East 200 South Salt Lake City, UT 84111 (801) 799-4165
	Dennis Montgomery Secretary	Ogden Fire Department 320 26th Street Ogden, UT 84401-3108 (80 1) 629-8074

International Association of Arson Investigators
CHAPTER PRESIDENTS AND SECRETARIES (Continued)

State	Contact	Address
Virginia	Frank A. Williams President	Virginia State Police P.O. Box 27472 Richmond, VA 23261 (804) 323-2433
	D. Darlene Greeny Secretary	3111 N. 17th Street Arlington, VA 22201 (703) 218-0519
Washington	Jack Avril President	11033 Dean Court, S.W. Tacoma, WA 98498 (206) 582-1505
	Ed Stokes Secretary	15902 63rd Avenue Court, E. Puyallup, WA 98373 (206) 563-4612
West Virgnia	Ken Johnson President	P.O. Box 1045 Beckley, WV 25802 (304) 256-2115
	Ken Shaffer Secretary	113 Lakeview Drive Charleston, WV 25313 (304) 776-3817
Wisconsin	Earl Meyer President	1097 Lori Lane Sun Prairie, WI 53590 (608) 837-8603
	Oscar Beilke, Jr. Secretary	2202 Hickory Lane New Holstein, WI 53061 (414) 849-2335
Wyoming	Dan Shatto President	240 Lincoln Lander, WY 82520 (307) 332-2870
	Roberta L. Smith Secretary	CCR Box E-4 Casper, WY 82637 (307) 237-2250

ORGANIZATIONS WITH SPECIALIZED ARSON CONCERNS

Alliance of American Insurers
1501 Woodfield Road
Suite 400 West
Schaumburg, IL 60173-4980
Contact: William R. Schroeder
(708) 330-8500

American Insurance Association
1130 Connecticut Avenue, N.W.
Suite 1000
Washington, DC 20036
Contact: Lawrence W. Zippin
Executive Vice President and Chief Operating
   Officer
(202) 828-7100
(202) 293-1219 (Fax)

American Insurance Services Group, Inc.
700 New Brunswick Avenue
Rannay, NJ 07065
Contact: Property Insurance Loss Register
(908) 388-0157

Association of Defense Trial Attorneys
124 S.W. Adams
Suite 600
Peoria, IL 61602
Contact: Gary M. Peplow
(309) 676-0400

Factory Mutual Engineering and Research
   Corporation
1151 Boston Providence Turnpike
P.O. Box 9102
Norwood, MA 02062
Contact: Paul M. Fitzgerald
(6 17) 762-4300

Fire Department Safety Officers Association
P.O. Box 149
Ashland, MA 01721-0149
Contact: Mary McCormack
Executive Director
(508) 881-3114
(508) 881-1128 (Fax)

Fire Marshal's Association of North America
One Batterymarch Park, P.O. Box 9101
Quincy, MA 02269-9101
Contact: Benjamin Roy
Executive Secretary
(617) 770-3000
(617) 770-0700 (Fax)

Insurance Committee for Arson Control
110 William Street
New York, NY 10038
Contact: Richard Gilman
Executive Director
(2 12) 669-9245
(212) 732-1916 (Fax)

Insurance Information Institute
110 William Street
New York, NY 10038
Contact: Gordon C. Stewart
P r e s i d e n t
(2 12) 669-9200

International Association of Arson Investigators
300 S. Broadway, Suite 100
St. Louis, MO 63102
Contact: Benny King
Executive Secretary
(314) 621-1966
(314) 621-5125 (Fax)

International Association of Chiefs of Police
Arson and Explosives Committee
555 N. Washington Street
Alexandria, VA 22314-2357
Contact: Carolyn Cockroft
Staff Liaison
(703) 836-6767

International Association of Fire Chiefs
4025 Fair Ridge Drive
Fairfax, VA 22033-9868
Contact: Garry L. Briese
Executive Director
(703) 273-0911

International Association of Fire Fighters
1750 New York Avenue, N.W.
Washington, DC 20006
Contact: Alfred K. Whitehead
President
(202) 737-8484
(202) 737-8418 (Fax)

International Fire Service Training Association
(IFSTA)
Fire Protection Publications
Oklahoma State University
Stillwater, OK 74078-0118
Contact: Gene Carlson
(405) 744-5723
(405) 744-8204

International Society of Fire Service Instructors
30 Main Street
Ashland, MA 01721
Contact: Edward McCormack
(508) 881-5800
(508) 881-6829 (Fax)

National Association of Mutual Insurance
Companies
P.O. Box 68700
3601 Vincennes Road
Indianapolis, IN 46268-0700
Contact: Dale Skupa
Senior Vice President
Government Affairs
(3 17) 875-5250
(317) 879-8408 (Fax)

National Association of State Fire Marshals
925 Madison Street
Jefferson City, MO 65101
Contact: John H. Coburn, Executive Director
(3 14) 636-4317
(314) 636-5262 (Fax)

National Association of State Foresters
Hall of the States, Suite 540
444 N. Capitol Street, N.W.
Washington, DC 20001
Contact: Paul D. Frey
(202) 624-5415
(202) 624-5407 (Fax)

National Crime Prevention Association
University of Louisville, Shelby Campus
Louisville, KY 40292
Contact: Wilbur Rykert
(502) 588-6787

The National Fire & Arson Report
P.O. Box 411087
Charlotte, NC 28241-1087
Contact: Barbara P. Goodnight
Publisher/Managing Editor
(800) 488-6327

National Fire Protection Association
One Batterymarch Park, P.O. Box 9101
Quincy, MA 02269-9101
Contact: George D. Miller
President
(615) 770-3000

National Insurance Crime Bureau
15 Franklin Street
Westport, CT 06880
Contact: Wendell C. Harness
(203) 226-6347

National Volunteer Fire Council
P.O. Box 25215
Alexandria, VA 22313-5215
Contact: Robert McKeon
Chairman
(203) 822-9411 (after 1 p.m.)

Property Loss Research Bureau
1501 Woodfield Road
Suite 400 West
Schaumburg, IL 60173-4978
Contact: Hugh O. Strawn
(708) 330-8650

Society of Fire Protection Engineers
One Liberty Square
Boston, MA 02109-4825
Contact: D. Peter Lund, CAE
Executive Director
(617) 482-0686
(617) 482-8184 (Fax)

Young Lawyers Division
American Bar Association
750 N. Lake Shore Drive
Chicago, IL 60611
Contact: Thomas H. Gonser
Executive Secretary
(3 12) 988-6235

PART III

Alphabetical Index

A

Adams, Claude, 87
Aetna Life and Casualty, 68
Aetna Life and Casualty Company, 63,64,66,67
Ahrens, Marty, 49
Alabama Arson Prevention Task Force, 6
Alabama State Fire College, 73
Alameda County Arson Task Force, 12
Alaska Fire Marshal Office, 47
Albany Arson Task Force, 32
Albany County Arson Task Force, 16
Albany Fire Department, 32
Albion Arson Task Force, 32
Alford, Ted J., 93
Alleghany County Task Force, 16
Alletto, William C., 31
Alliance of American Insurers, 68,99
Allstate Insurance Company, 67,68
Amabili, Louis J., 74
American Insurance Association, 99
American Insurance Services Group, Inc., 99
Amsterdam Arson Task Force, 33
Amsterdam Fire Department, 33
Anderson, John, 81
Anderson, Lloyd, 8
Anthony, Virginia, 12
Anti-Arson Committee of Connecticut Fair Plan, 6
Arkansas Arson Advisory Committee, 6
Arkansas Fire Academy, 47,73
Arnold, David W., 86
Arson Advisory Board, 7
Arson Alarm Foundation, 11
Arson Alert Program of Iowa, 8
Arson Control Association of Maryland, 8
Arson Strike Force, 30
Asheville-Buncombe Arson Task Force, 24
Association of Defense Trial Attorneys, 99
Atchley, William, 14
Atlanta Fire Department, 3 1
Atlantic County Prosecutor's Office, 15
Auburn Arson Task Force, 33
Aurora Fire Department, 54
Avril, Jack, 98

B

Babcock, Wayne, 81
Baca, Bob, 78
Baccari, Marie, 60,64,66
Bailey, Steven C., 56

Baker, Byron J., 22
Ballard, Jack, 30
Barbaro, Frank, 54
Barber, Keith M., 16
Barker, Mark, 73
Barnes, Donald, 21
Baron, Dennis M., 34
Barracato, John, 85
Barry, James, 37
Barry, Mike, 91
Bartig, Gerald, 87
Bartow, David M., 20
Batavia Arson Task Force, 33
Baton Rouge Fire Department Arson Division, 32
Baughn, W. "Ronnie" T., 92
Beachcamp, Charles, 92
Beck, Sue, 89
Beilke, Oscar, Jr., 28, 98
Bell, A.D., 47
BellingharnWbatcom County Arson Task Force, 27
Belstner, Leonard H. "Skip," 11
Benton, E.E., 43
Benz, Karl J., 90
Bergen County Prosecutor's Office, 15
Bergin, Edward D., 30
Berry, Gene, 55
Binghamton Arson Task Force, 33
Binghamton Fire Department, 33
Bliss, Donald P., 49
Bliss, Rosemary, 13
Blowers, Ross, 94
Boeh, Dan, 89
Bolis, Dan, 79
Bond, Richard, 24
Borden, James, 38
Boston Fire Department, 54
Brace, Thomas R., 49
Bradley, Tim, 50
Bradshaw, Tom R., 42
Bradt, Frank A., 18
Bridgeport Arson Task Force, 30
Briese, Garry L., 100
Bright, Lee, 48
B&ton, Allen W., 93
Brizzcll, Jim, 50
Brooks, David J., 81
Broome County Arson Task Force, 16
Brown County Fire Investigation Task Force, 28
Brown, Gerald R., 97
Brown, Rodney, 87
Buffalo Arson Task *Force, 34*

www.ingramcontent.com/pod-product-compliance
Lightning Source LLC
Chambersburg PA
CBHW081132170526

45165CB00008B/2645